2011-2012

ASSESSMENT OF THE ARMY RESEARCH LABORATORY

Army Research Laboratory Technical Assessment Board

Laboratory Assessments Board

Division on Engineering and Physical Sciences

NATIONAL RESEARCH COUNCIL
OF THE NATIONAL ACADEMIES

THE NATIONAL ACADEMIES PRESS
Washington, D.C.
www.nap.edu

THE NATIONAL ACADEMIES PRESS **500 Fifth Street, NW** **Washington, DC 20001**

NOTICE: The project that is the subject of this report was approved by the Governing Board of the National Research Council, whose members are drawn from the councils of the National Academy of Sciences, the National Academy of Engineering, and the Institute of Medicine. The members of the authoring board responsible for the report were chosen for their special competences and with regard for appropriate balance.

This study was supported by Contract No. DAAD17-03-C-0017 between the National Academy of Sciences and the Army Research Laboratory. Any opinions, findings, conclusions, or recommendations expressed in this publication are those of the authors and do not necessarily reflect the views of the agency that provided support for the project.

International Standard Book Number-13: 978-0-309-26899-8
International Standard Book Number-10: 0-309-26899-0

Copies of this report are available from

Laboratory Assessments Board
Division on Engineering and Physical Sciences
National Research Council
500 Fifth Street, NW
Washington, DC 20001

Additional copies of this report are available from the National Academies Press, 500 Fifth Street, NW, Keck 360, Washington, DC 20001; (800) 624-6242 or (202) 334-3313; Internet http://www.nap.edu.

Printed in the United States of America

THE NATIONAL ACADEMIES

Advisers to the Nation on Science, Engineering, and Medicine

The **National Academy of Sciences** is a private, nonprofit, self-perpetuating society of distinguished scholars engaged in scientific and engineering research, dedicated to the furtherance of science and technology and to their use for the general welfare. Upon the authority of the charter granted to it by the Congress in 1863, the Academy has a mandate that requires it to advise the federal government on scientific and technical matters. Dr. Ralph J. Cicerone is president of the National Academy of Sciences.

The **National Academy of Engineering** was established in 1964, under the charter of the National Academy of Sciences, as a parallel organization of outstanding engineers. It is autonomous in its administration and in the selection of its members, sharing with the National Academy of Sciences the responsibility for advising the federal government. The National Academy of Engineering also sponsors engineering programs aimed at meeting national needs, encourages education and research, and recognizes the superior achievements of engineers. Dr. Charles M. Vest is president of the National Academy of Engineering.

The **Institute of Medicine** was established in 1970 by the National Academy of Sciences to secure the services of eminent members of appropriate professions in the examination of policy matters pertaining to the health of the public. The Institute acts under the responsibility given to the National Academy of Sciences by its congressional charter to be an adviser to the federal government and, upon its own initiative, to identify issues of medical care, research, and education. Dr. Harvey V. Fineberg is president of the Institute of Medicine.

The **National Research Council** was organized by the National Academy of Sciences in 1916 to associate the broad community of science and technology with the Academy's purposes of furthering knowledge and advising the federal government. Functioning in accordance with general policies determined by the Academy, the Council has become the principal operating agency of both the National Academy of Sciences and the National Academy of Engineering in providing services to the government, the public, and the scientific and engineering communities. The Council is administered jointly by both Academies and the Institute of Medicine. Dr. Ralph J. Cicerone and Dr. Charles M. Vest are chair and vice chair, respectively, of the National Research Council.

www.national-academies.org

ARMY RESEARCH LABORATORY TECHNICAL ASSESSMENT BOARD

LYLE H. SCHWARTZ, NAE,[1] Air Force Office of Scientific Research (retired), *Chair*
DONALD M. CHIARULLI, University of Pittsburgh
DAVID E. CROW, NAE, Pratt and Whitney Aircraft Engine Company (retired)
MARJORIE ERICKSON, Phoenix Engineering Associates, Inc.
DEBASIS MITRA, NAE, Bell Labs, Alcatel-Lucent
R. BYRON PIPES, NAE, Purdue University
JEREMY M. WOLFE, Brigham and Women's Hospital and Harvard Medical School

Staff

JAMES P. McGEE, Director
ARUL MOZHI, Senior Program Officer
LIZA HAMILTON, Administrative Coordinator
EVA LABRE, Program Associate
ROSE NEUGROSCHEL, Research Associate (until January 20, 2012)

[1]NAE, membership in the National Academy of Engineering.

Acknowledgment of Reviewers

This report has been reviewed in draft form by individuals chosen for their diverse perspectives and technical expertise, in accordance with procedures approved by the National Research Council's Report Review Committee. The purpose of this independent review is to provide candid and critical comments that will assist the institution in making its published report as sound as possible and to ensure that the report meets institutional standards for objectivity, evidence, and responsiveness to the study charge. The review comments and draft manuscript remain confidential to protect the integrity of the deliberative process. We wish to thank the following individuals for their review of this report:

Claude R. Canizares, Massachusetts Institute of Technology,
Don B. Chaffin, University of Michigan,
George E. Dieter, University of Maryland,
Michael Dunn, Ohio State University,
David R. Ferguson, Issaquah, Washington,
William D. Gropp, University of Illinois at Urbana-Champaign, and
Edmund O. Schweitzer III, Schweitzer Engineering Laboratories, Inc.

Although the reviewers listed above have provided many constructive comments and suggestions, they were not asked to endorse the conclusions or recommendations, nor did they see the final draft of the report before its release. The review of this report was overseen by Joseph S. Francisco, Purdue University. Appointed by the National Research Council, he was responsible for making certain that an independent examination of this report was carried out in accordance with institutional procedures and that all review comments were carefully considered. Responsibility for the final content of this report rests entirely with the authoring board and the institution.

Contents

APPENDIXES

Summary

The charge of the Army Research Laboratory Technical Assessment Board (ARLTAB) is to provide biennial assessments of the scientific and technical quality of the research, development, and analysis programs at the Army Research Laboratory (ARL). The ARLTAB is assisted by six panels, each of which focuses on the portion of the ARL program conducted by one of ARL's six directorates.[1] When requested to do so by ARL, the ARLTAB also examines work that cuts across the directorates. For example, during 2011-2012, ARL requested that the ARLTAB examine crosscutting work in the areas of autonomous systems and network science.

The overall quality of ARL's technical staff and their work continues to be impressive. Staff continue to demonstrate clear, passionate mindfulness of the importance of transitioning technology to support immediate and longer-term Army needs. Their involvement with the wider scientific and engineering community continues to expand. Such continued involvement and collaboration are fundamentally important for ARL's scientific and technical activities and need to include the essential elements of peer review and interaction through publications and travel to attend professional meetings, including international professional meetings. In general, ARL is working very well within an appropriate research and development niche and has been demonstrating significant accomplishments, as exemplified in the following discussion, which also addresses opportunities and challenges.

The recently initiated multiscale modeling Collaborative Research Alliances (CRAs) are ambitious yet necessary efforts. ARL is to be commended for establishing the two CRAs, the Multiscale Multidisciplinary Modeling of Electronic Materials (MSME) and the Materials in Extreme Dynamic Environments

[1]The six ARL directorates are the Computational and Information Sciences Directorate (CISD), Human Research and Engineering Directorate (HRED), Sensors and Electron Devices Directorate (SEDD), Survivability/Lethality Analysis Directorate (SLAD), Vehicle Technology Directorate (VTD), and Weapons and Materials Research Directorate (WMRD). Appendix A provides information summarizing the organization and resources of ARL and its directorates.

(MEDE). Under the MSME, a multi-university team is charged with doing the fundamental multiscale modeling on lithium batteries, fuel cells, and electronic materials and devices. The supporting experimental work will be done at ARL. Given the substantial issues concerning the premise of bridging the scales in multiscale modeling, many challenges (and thus opportunities) exist in demonstrating verifiable success in the time frame of 5 years. Close interactions of ARL personnel with the multi-university team selected will be critical to ensure a focused and productive group effort.

BATTLEFIELD ENVIRONMENT DIVISION AND COMPUTATIONAL SCIENCES DIVISION OF THE COMPUTATIONAL AND INFORMATION SCIENCES DIRECTORATE

The Army has a unique and pressing requirement for near-Earth atmospheric understanding and characterization beyond what can be provided by other military and civilian entities. The Battlefield Environment Division (BED) has been responding to this need by addressing fundamental scientific problems and is making progress toward becoming a first-class research organization. BED's work in the following areas involves creative combinations of theory and experimentation and of hardware and software to move toward solutions of challenging problems: turbulence propagation theory and effects, weather research and forecast model-based nowcasting for battlefield operations, systems aimed at single-particle detection for use in biohazard threat applications, and optical systems for atmospheric sensing. Continued efforts to validate underlying models and demonstrate practical applications for these and other projects are necessary and planned. Consistent, appropriate verification of the performance of mesoscale and microscale models remains a challenge to the wider meteorological numerical modeling community, and BED is still at an early stage in arriving at an assessment scheme for its various numerical prediction models.

The overall technical quality of the research conducted in the Computational Sciences Division (CSD) is improving. Some CSD projects are of very high quality. For example, the multiscale materials modeling work involves ambitious and powerful goals that build on ARL strengths. The work demonstrates a very good basic science approach that supports an important and large ARL enterprise in multiscale modeling. The division's multiscale materials modeling program requires intensive computational capabilities. There was evidence that CSD understands that verification and validation are required not just for the models, but for the data exchanges between the models and transitions between scales.

The structure of the CSD includes a substantial facilities component that serves the high-performance computing and networking infrastructure needs of ARL, the Army, and the Department of Defense (DoD); now it also includes a growing research component focused on interdisciplinary computational science. The CSD team has made substantial progress in a very short time in articulating a research vision and realigning activities to support that vision within the research component.

Within CSD, awareness of prior extramural work continues to be uneven. To confirm this awareness and the technical quality of its work, CSD should employ the mechanisms available in academic circles—for example, peer-reviewed publication and attendance at conferences, symposia, and other professional meetings.

HUMAN RESEARCH AND ENGINEERING DIRECTORATE

Research in the Human Research and Engineering Directorate (HRED) is currently organized around eight research thrusts—sensory performance, physical and cognitive performance interaction, translational neuroscience, social-cognitive network science, human-robot interaction (HRI), human-systems integration (HSI), opportunity-driven human factors research, and simulation and training technology.

In this Summary the social-cognitive network science thrust and the HRI thrust are discussed within the sections below titled "Network Science Enterprise" and "Autonomous Systems," respectively. Work performed in association with the other thrusts is discussed in this section.

In general, the strongest work in the area of sensory performance is the customer-driven evaluation of equipment. The overall research program in this area has lacked clear direction, although progress is visible. The group is not yet a force in basic or applied research on sensory performance. Recently, a restatement of this area's vision has appeared under the rubric of "owning the environment." Under that rubric, the emphasis is on the use of stealth and deception as force multipliers. That work is in an early stage and has not yet given rise to a program of research.

The research on physical and cognitive performance interaction is of generally high technical quality. Its scope has been somewhat limited. However, with the recent development of new facilities, this may be viewed as the start of a successful, coherent program of research. Biomechanics has long been at the core of HRED's mission, and the directorate has developed a very strong set of facilities for studying soldier performance under various approximations of real-world situations. With these facilities, HRED has identified a niche that it is perhaps uniquely suited to study—the interaction of physical and cognitive stressors on performance.

Supported by a cadre of university experts in a Collaborative Technology Alliance (CTA), the ARL work in translational neuroscience is indicative of high-quality neuroscience research that is routinely validated by its publication in good, peer-reviewed journals. Over the past 5 years, ARL's neuroscience group has grown into the DoD's largest internal nonmedical translational neuroscience research effort, and it continues its trajectory toward being a neuroscience laboratory on par with strong university research programs. Work on detection and classification of artifacts in electroencephalogram (EEG) signals is noteworthy.

During the past 2 years, the social-cognitive network science group has worked to refine its vision and to align its work with the network science work in the Computational and Information Sciences Directorate (CISD). The overall quality of the research in this area is uneven. There is some fine work, but also much work that will have low impact, either because it is methodologically weak or because it does not appear to be part of a systematic program of work. Addressing the challenge of network science may be an opportunity to bring in staff with different backgrounds from those of current HRED staff.

HSI represents one of the core HRED competencies: providing the Army with assessments of how humans will work with new systems. Much of the work is very solid, high-quality human factors work, employing a range of tools, including the IMPRINT (Improved Performance Research Integration Tool) model of human performance. IMPRINT is the crown jewel of HRED's HSI efforts. Its continued development and use represent a strength of the program. It is possible that there is a gap in IMPRINT in the form of inadequate coverage of cognition and perception. If so, given the nature of today's Army tasks, this gap should be addressed. The use of HRED's HSI expertise can also be seen in work such as that on progressive insertion of human figure modeling into the acquisition process.

Cases in which no previous research exists for providing a solution to a problem produce the opportunity for research. The HRED "opportunity-driven" research projects presented were generally of high quality. This nicely illustrates the ability of HRED to serve as a problem-solving resource across a wide range of Army problems. Whereas the HSI effort is targeted toward intervening early in the acquisition process, MANPRINT (Manpower and Personnel Integration; the Army's program for considering manpower and personnel factors during acquisition) equipment design issues can be identified during usability testing or through reports received from the field. This may also be a place where scientific mentoring by senior researchers would be helpful. The data being gathered by the opportunity-driven research represents an opportunity to close the loop between basic and applied research and soldier system problem solving. To support this effort, the hiring of more senior, experienced scientists should be complemented by the better use of available field operations reports, and even medical reports, to prioritize studies and plan how they should be executed.

The overall quality of the work reviewed in the simulation and training technology area of HRED appears to be good. The pace and diversity of Army missions require a rapid, responsive training capability. The efforts at the Simulation and Training Technology Center (STTC) appear to represent a generally impressive program of work, although comparatively few program details were presented to the ARLTAB's review panel. The evaluation reported here is somewhat tentative, because the ARLTAB has not seen the facilities or met many of the staff. Staff at the STTC have made impressive efforts to link STTC work to broader work in the field.

Within an overall program of solid research in HRED, a number of projects appear to be deficient in conception and/or execution. Causes of difficulty include inadequate attention to relevant literature, lack of coherent supporting theory, and inadequate analysis of results. Some HRED studies appear to be statistically underpowered, and some statistical analyses have been questionable. A more subtle issue is the distinction between statistical significance and scientific or practical significance. It is possible to have statistically reliable results that are, nevertheless, of little consequence. It was not always clear that this distinction was recognized by some HRED researchers.

SENSORS AND ELECTRON DEVICES DIRECTORATE

The quality of research being conducted within the Sensors and Electron Devices Directorate (SEDD) is of highest caliber. Considering the numerous awards received by the SEDD scientific staff, the number of refereed publications authored and co-authored, and the number of presentations given at professional society meetings, the staff compares favorably with that of other industrial and university research institutions. From a technical viewpoint, SEDD is accomplishing well the enormous task of understanding the needs of an astonishingly broad range of Army applications and then providing innovative solutions. It is impressive that, in this context, SEDD endeavors to predict Army needs that are one to two decades in the future. For example, SEDD has quickly become a leader in the scientific community working in the nascent field of extreme energy science. SEDD also plays a strong role in the ARL Network Science Enterprise and in the ARL Multiscale Modeling Enterprise.

In the area of micropower, the notion of a fully integrated (electronics and passives), single-chip power supply is an elusive goal in the power electronics business. SEDD researchers have made important progress, producing results that are among the best in the field. Significant progress has been made in the area of fuel cells. In particular, a field test of the M-100 direct methanol fuel cell system demonstrated a high energy density that will reduce logistics and provide both cost savings and weight savings. SEDD researchers have demonstrated excellent voltage stability of lithium-ion (Li-ion) cells. SEDD continues to be the leader in research on silicon carbide (SiC) electron devices. SEDD has demonstrated a resonant 40 kV, toward 120 kV, DC/DC converter that uses SiC's advantages of high voltage and perhaps higher temperature capabilities to meet SWaP (size, weight, and power) requirements. In other power electronics efforts, the implementation of SiC devices from large-diameter wafers for DC through pulsed applications is a significant achievement.

The work on high-performance and high-value passive circuit elements has significantly advanced the state of the art. In particular, SEDD has demonstrated miniature on-chip inductors fabricated by thick electroplating, with high quality factors and high inductance density. SEDD has developed advanced design techniques for ultrawideband digital-to-analog converters (DACs) and has simulated an 8-bit, 32 GS/s DAC. SEDD maintains a thriving microelectromechanical systems (MEMS) capability.

Accomplishments range from fundamental materials-development efforts in graphene and piezoelectrics to near-term devices such as the traumatic brain injury sensor. SEDD researchers have achieved an increase in the power efficiency of an InAs/GaAs quantum dot solar cell from 9 to 14 percent. Resonator quantum-well infrared photodetectors (QWIPs) continue be a crown jewel among the achievements of SEDD, which has been able to incorporate advanced optical concepts into the detector design. SEDD has developed a better understanding of new near-field optical phenomena that has enabled the design of a QWIP structure with quantum efficiencies nearly 70 percent, doubling the previous record of 35 percent for corrugated-QWIPs.

Research aimed at creating extremely low-noise oscillators has produced outstanding results. SEDD has found a way to build a 10 GHz optoelectronic oscillator that matches the specifications of commercial low-noise oscillators up to 10 kHz away from the carrier for less than an order-of-magnitude lower cost and competitive size and weight specifications. The temperature achieved in a laboratory demonstration of cold atom optics technology was a remarkable 40×10^{-6} K. SEDD researchers, in collaboration with the University of Central Florida, have demonstrated high-quality AlGaN/Ga:MgZnO grown on sapphire.

Under the Multiscale Multidisciplinary Modeling of Electronic Materials Collaborative Research Alliance, a multi-university team is charged with doing the fundamental multiscale modeling of lithium batteries, fuel cells, and electronic materials and devices. The supporting experimental work will be done at ARL. Close interactions of ARL personnel with the multi-university team selected will be critical to ensure a focused and productive group effort. A potential challenge is to make sure that the multi-university team, with principal investigators having different (not necessarily overlapping) skills, remains focused and delivers value to ARL.

The rotating SEDD directorship is not healthy for the organization. Although talented division chiefs have fulfilled this role at 4-month, rotating intervals, these short stints of leadership do not allow for sustained and consistent assessment of how the needs of the customers match the current research and development (R&D) activities, for the alignment of funding and personnel resources in the light of shifting missions and budgets, and for the tweaking of the constraints to adjust the flow of innovation. A permanent director is needed for SEDD.

SURVIVABILITY/LETHALITY ANALYSIS DIRECTORATE

The Survivability/Lethality Analysis Directorate (SLAD) aims to provide sound assessment and evaluation support of the survivability, lethality, and vulnerability (SLV) of Army equipment and soldier systems. SLAD facilities include state-of-the-art laboratories and equipment in which to conduct research and which provide a strong basis for potential collaborations with outside partners. SLAD has many opportunities to capitalize on its testing and evaluation base so as to expand programs selectively, to develop tools and methodologies to broaden its analysis capabilities, and to define and maintain the competitive edge required for SLAD to be the Army's primary source for SLV assessment.

SLAD's collaborations with other ARL directorates continue to improve. For example, a collaborative effort between SLAD and HRED is built on the expertise of each organization to rapidly develop a new anthropomorphic test device—a warrior injury assessment manikin (WIAMan)—for assessing injury from underbody blast. SLAD researchers also collaborated with academic researchers to obtain better data by using cadavers instead of anthropomorphic test devices.

A new program to develop a metric to predict mild traumatic brain injury resulted in the design of a surrogate sensor system that mimics the effect on the human brain of blast and blunt trauma. A process for rapid determination of the level of trauma in the field has been proposed as well.

The Integrated Network Vulnerability Assessment/ Discovery Exploitation (INVA/DE) tool is excellent for detecting the vulnerability of computer network operations. Further collaborations with CISD to pit SLAD's INVA/DE against CISD's Interrogator will test the robustness of each program. This impressive program seems very well tied into the intelligence community.

The ballistic analysis of a lightweight vehicle-protection system demonstrates the value of SLAD analysis and testing in supporting Army acquisitions. The active protection systems program has provided new results on the residual threat of light-armored vehicles from rocket-propelled grenades.

Under SLAD's direction, contractors at New Mexico State University continue the development of the System-of-Systems Survivability Simulation (S4) software code. After a long period of initial development, progress has begun toward establishing a formal software development methodology and in bringing the system-of-systems analysis program, including S4, more under SLAD control. SLAD remains far from establishing itself as a credible participant in the DoD system-of-systems analysis environment. Currently, the supporting facilities are state-of-the-art, but the program does not reflect a broad understanding of the underlying science; the qualifications of team members are inadequate for the task; the analytical work does not reflect a sound understanding of Army requirements; and the mix of theory and computation is appropriate, but the theoretical basis is inadequate.

SLAD should consider a tactical pause to review and more carefully define the role and mission of S4, carefully reexamine its technical plan for collaborating with and supporting the efforts of other modeling activities within the DoD (such as the Army Training and Doctrine Command Analysis Center [TRAC]), focus its system-of-systems analysis modeling on cases that demonstrate the usefulness of S4 to others, and align its resources with its plan. The S4 team should investigate integration of the ACQUIRE target acquisition model to augment the S4 sensor models, develop a consolidated set of new communication features for S4 with input from the Army communications community, develop a solid engagement with the intelligence community, structure S4 to accept combat dynamics as an input from other Army combat models, and leverage the services of the Army Materiel Systems Analysis Activity (AMSAA) for validation, verification, and assessment. SLAD should focus on physics processes at a high-fidelity level (e.g., packet-level resolution in communications).

VEHICLE TECHNOLOGY DIRECTORATE

Spurred by the base realignment and closure process that consolidated the Vehicle Technology Directorate (VTD) at Aberdeen Proving Ground in Maryland, VTD's realigned focus on Army needs, such as ground vehicle technology and autonomous systems, has been increasing the quality of its research portfolio. In 2010 the establishment of eight capability concepts that embody the technical breakthroughs needed to meet critical future Army needs was a major step toward focusing and upgrading VTD research. However, the failure of the directorate in the period of time covered by this report to continue to align the VTD research portfolio with these capability concepts represents a significant lost opportunity. It is critical that VTD quickly refocus on the technologies needed for the capability concepts. For example, it was recognized in 2010 that a crawling-bug type vehicle needed to be added to the microautonomous systems portfolio of research, but it has not yet been added. Similarly, combustion of JP-8 fuel in the small volume of small engines is a technology area that would impact several capability concepts and is therefore a high-priority, but unaddressed, research area. Without a systematic plan to meet critical Army needs, the quality of VTD's research will continue to degrade.

VTD's most significant accomplishments over the past 2 years are in the areas of the Robotics CTA and the Micro Autonomous Systems and Technology (MAST) CTA and the start-up and checkout of the new VTD laboratory areas.

The project on continuous trailing-edge flap for helicopter rotors evinced the highest quality among the VTD projects presented; it is of the caliber of the best in its field. This is a well-conceived and well-designed program using a novel approach—the novelty being to use a piezoelectric actuated biomorph embedded in a rotor blade for the purpose of trailing-edge deflection. The concept offers an alternative approach to discrete flaps, commonly being researched for application to helicopter rotor blades.

Projects such as physics of failure from multidynamic excitation, separated flowing using nanosecond pulsed plasma, and ducted rotor research contain important aspects but are not of the highest caliber. Other projects, such as material damage precursors in composite structures, variable-speed power turbine research, compressive sensing robust recovery of sparse mechanical signals from incomplete measurements, and slowed-rotor unpowered take-off, are of inadequate quality.

The Small Engine Altitude Performance and Heat Engine Systems Altitude Test Facility is state of the art, and it is unique relative to facilities of other government laboratories. The facility has the capability to simulate sea level to 25,000 feet, can provide air from –40 °F to +130 °F, and has two dynamometers. These dynamometers together cover a span from 1 to 250 horsepower (hp). The current work is focused on small engines of 40 hp or less under a subatmospheric, variable-pressure environment in which engine performance and components will be evaluated.

The Army's objective of utilizing JP-8 fuel in all of its vehicles makes the combustion facility a necessity for VTD. The utilization of JP-8 over the range of engines of interest to the Army is a formidable challenge; it is not clear that the full depth of this problem is understood by VTD. Perhaps VTD should also add spark assist and catalytic breakdown of the JP-8 to its portfolio in order to help attack this difficult problem. The current combustion facility allows for some interesting experiments; however, the facility needs to be modified to cover the entire range of sizes and combustion configurations of interest to the Army. VTD should consider how all of the concepts discussed, such as flash heating of the fuel, fit into an overall program aimed at delivering JP-8 combustion over a range of engine varieties and sizes.

Many of the Army's needs involve air and ground vehicles at relatively low speeds, and so the low speed wind tunnel fills a VTD requirement. However, plans to test the hover flight conditions and the slow forward flight of micro air vehicles in this wind tunnel are likely to yield unrealistic results. The utilization of a free jet may be required to test the micro air vehicles accurately. ARL needs to purchase the velocimeter equipment necessary to fully check out and test vehicles in the wind tunnel. The turbulence level in the test section of the tunnel has not been measured. VTD management should obtain help from outside experts before undertaking this effort.

WEAPONS AND MATERIALS RESEARCH DIRECTORATE

The overall scientific quality of the work at the Weapons and Materials Research Directorate (WMRD) is comparable to that of other national laboratories. Generally, the WMRD programs reflect a broad understanding of the science and engineering underlying their work. WMRD has appropriate laboratory equipment and numerical codes and models. In particular, the shock physics laboratories are impressive. Computational tools are being used extensively, but there could be more characterization equipment, and in some cases WMRD could adopt additional numerical codes and models from outside ARL. The qualifications of the research staff are excellent, and their knowledge is generally very solid. WMRD's work reflects understanding of the Army's needs—an example is the development and fielding of the new bullet. WMRD applies to its research an appropriate mix of theory, computation, and experimentation.

WMRD presented examples of quality research programs that allowed novel investigations leading to new discovery or to applications beyond the initial intent of programs. For example, the study that characterized ceramic microstructure by means of capacitance measurement and then correlated to ballistic measures is one clear example of creative thinking, because it recognized the statistical characteristics of the measurements and cleverly used Bayesian methods to extract dominant ballistic behavior.

Efforts in materials chemistry reflect high-quality work spanning science and engineering. The use of combinatorial chemistry and property modeling of the behavior of individual molecules uses state-of-the-art chemical selection pioneered by the pharmaceutical industry. At the other extreme, off-the-shelf materials with careful selection of ligands and surface activity are used to develop cost-effective coatings that can be adapted rapidly to a wide range of specific threats. In the future, the merging of the combinatorial chemistry and science with the surface engineering of coatings should provide a strong foundation for important contributions.

Many of WMRD's armor technology efforts are impressive. For example, the kinetic energy (KE) armor technology work illustrates how a back-to-basics approach can provide a significant, long-term payoff for the Army. The experimental results were combined with multidimensional computational modeling in order to gain a better understanding of how the defeat mechanism worked and how the armor technology could potentially be improved or optimized. This eventually led to a multiyear, applied R&D effort focused on further maturing and demonstrating this new KE armor technology in practical armor designs. This work has recently resulted in full-scale prototype KE armors that show significant tactical potential for use in ground combat vehicles and other applications.

In WMRD there is a developing continuum of work, coupled to Army-specific needs, from basic research through applications. One notable example is the work in new forms of energetic materials. The specific work on extended solids and nanodiamonds for structural bond energy release is technically high-risk and potentially high-payoff work. The example of the 885A1 system successes in lethality and in the reduction of environmentally sensitive materials is noteworthy as well. Deployment of the system without complete laboratory validation has proved to be a wise and timely choice.

WMRD's cold spray capabilities continue to show great promise for applications. This work has demonstrated that reactive materials can be used to tailor the lethality profile of the fragments as a function of distance with a density high enough to compete with conventional inert fragments (7 g/cc). It was found that processing challenges exist in making homogeneous materials using cold spray. Composite particles could possibly result in more homogeneous samples, as well as in tunable reactivity.

The high-pressure polymerization of CO and N is work representing leading-edge research in the polymer field. It is important to characterize Poly Co and Poly N as polymers with properties such as molecular weight and glass transition temperature. The goal of producing 1 gram of polymer this year may not represent enough to characterize these polymers. These characterizations are important for validating the computational data on the various species being predicted.

ARL is seeking to develop the capability to design, optimize, and fabricate lightweight protection material systems exhibiting revolutionary performance. The approach is to realize a materials-by-design capability by establishing the new CRA focused on materials in extreme dynamic environments. The focus of the CRA is to advance the fundamental understanding of materials relevant to high-strain-rate and high-stress regimes. The CRA is intended to create a collaborative environment that enables an alliance of participants from academia, government, and, potentially, industry and/or nonprofit organizations to advance the state of the art and to assist with the transition of research in order to enhance the performance of materials of interest to the Army. This effort is clearly a major undertaking that, to ensure success, will likely require a new approach and forms of technical management both within ARL and in the multi-university consortium.

NETWORK SCIENCE ENTERPRISE

The Network Science Enterprise is spearheaded by CISD's Network Sciences Division (NSD) and Information Sciences Division (ISD), but it also involves other ARL directorates and external collaborators participating in the Network Science Collaborative Technology Alliance and the International Technology Alliance (ITA). The enterprise is a work in progress, but it appears promising and pointed in the correct direction. The enterprise concept involves the high aspiration of integrating and coordinating efforts by many researchers who address multiple network types, termed genres (social-cognitive, information, and communications) that span a very broad spectrum of activities with shared networks, linkages, and dependencies.

ARL's vision of abstracting common concepts and mathematical structures across genres is laudable. The challenges associated with such combining are large. The difficulties of working across intellectual cultures, disciplines, technologies, and timescales are notorious. By the same token, when such combining does work, the results are often spectacular. To its credit, ARL has taken on the challenge. ARL is one of the pioneers among major laboratories in the world in committing resources and building a program dedicated to network science on the scale that it has.

There have been several achievements to date. A White House press release of March 14, 2012, recognized the work of ARL, noting the collaborative research done by the United States and United Kingdom as ITA partners to enhance information sharing and distributed, secure, and flexible decision making in coalition operations. Solid work on optical communications in networks is aimed at developing unconventional optical communication systems, including ultraviolet non-line-of-sight and covert visible-light communications, for applications that include intraconvoy communications in situations where jammers are being used to affect communications devices. The studies of social-cognitive networks involve a broad set of topics related to the way in which soldiers interact with networks; this work holds great promise but will require careful collaboration among social scientists, computer scientists, and engineers. It will also require the definition of a program of research that reflects the Army niche within the broad community of such researchers.

Two additional projects provide information that can represent key inputs to networked systems. The project in machine translation of text, supported by effective interaction with extramural researchers, continues to expand the capabilities of devices used in the field to translate foreign documents. The project on applied anomaly detection brings together experts in cognitive processing, machine learning, sensors, and military operations to develop means of training soldiers to sense dangerous situations in the field.

ARL should consider devoting more attention to two areas of research: cloud computing and cybersecurity. Tactical applications of cloud computing, such as tactical cloudlets, are particularly demanding variants that should be tackled only after meeting the prerequisite of adequate mastery of the science and technology of basic cloud computing. ARL should consider making available Hadoop clusters to jump-start research from the bottom up. These clusters, supported by the necessary software engineering knowledge, should be used as platforms to develop machine learning, data mining, search, social networks analysis, and other applications. Other dimensions to cloud computing, especially for tactical applications, call out for collaborative research. Examples include cybersecurity and special communications network protocols and resource scheduling to handle near-real-time applications within the geographically dispersed context of cloud computing.

CISD lists cyberdefense as one of its "top five future big ideas" and is developing a strategic approach to research in this area. There are many research questions related to advances in data leak prevention that require new insights and technical developments not shackled by the biases of unimaginative approaches to authentication and authorization. ARL may wish to consider work in data leak prevention. Scalable deception is an underdeveloped technology that is directly responsive to data leak prevention and is fraught with challenging research problems. Also worthy of further consideration are the topics science of security and security metrics. These, together with the topic of insider threat, pose research problems that a small number of high-quality researchers in ARL can address with potentially high impact.

ARL has devised a strategy for the Network Science Enterprise that has the following main elements: science driven by operational experience, expanding collaborations and partnerships, and a focus on experimentation. This is a good start. The strategy needs to be fleshed out in greater detail and documented. There is evidence of promising starts in the enterprise approach and in expanding collaborations and partnerships, notably in the partnerships with universities in the CTA and the ITA.

ARL is deriving value from the collaborative alliances. There is undoubted value from being aware and being involved, at any level, in high-quality research. However, to be confident that the collaborations will produce the impact that ARL desires, ARL should perform an in-depth investigation of the means and desired outcomes for extracting value and desired impacts from these collaborations. Such an investigation should consider two models for extracting value from such collaborations. One model assumes that ARL researchers interact on equal intellectual terms with their university collaborators. The second model assumes that the role of ARL staff is to instruct university collaborators in ways that will direct and transition research to address Army needs. In either case, ARL researchers will need sufficient intellectual depth to interact sufficiently on all projects with their university counterparts.

AUTONOMOUS SYSTEMS ENTERPRISE

ARL's most important accomplishment in the area of robotics has been the use of the MAST CTA and Robotics CTA to leverage high-caliber research and talent across the United States. Fully autonomous robotic operation may take 30 years to develop.

Many, but not all, of the research projects in the Autonomous Systems Enterprise are of the highest caliber; the combined quality of the research contained in the CTAs is cutting edge. The overall technical quality of the work is very high for each of the key areas addressed: microelectronics, sensing, signal processing, and perception; intelligence; HRI; and manipulation and mobility. The scientific quality of the research is comparable to that executed at federal, university, and/or national laboratories both nationally and internationally. The overall research reflects a broad understanding of the underlying science and research conducted abroad. Appropriate laboratory equipment and models are being used. The qualifications of the research team are very good. The facilities and laboratory equipment are state of the art.

Overall, the research portfolio in the area of microelectronics, sensing, signal processing, and perception covers well the size range of robots. The MAST program is doing best-in-class work at reducing sensor size and weight. As robots become smaller, the ability to carry sensors and support large computing activity decreases; there is a need to address scalability challenges with regard to capabilities (for example, fidelity of the sensors, range, and detection). Off-ramps to divert developed technology into existing platforms should receive more emphasis.

The selection of sensors to accomplish the perception element of autonomous systems should be driven by the task and mission to be accomplished and the objects, activities, and events that one is trying to find and characterize. This approach leads to the identification of what can be observed and a selection of the sensor suite and measurements associated with the set of observables. Little discussion was provided to the ARLTAB concerning the observables needed and the justification for the selection of sensors and associated processing.

The work on robotic intelligence showed slow incremental progress. Because this area is so difficult, ARL has appropriately been applying many approaches to solving the problems involved. However, many of these approaches and efforts are not in full coordination with others in the same area, and so researchers have not been able to leverage the lessons learned and emerging results of the other efforts. Relevant robotics research in the civilian market continues to dwarf ARL's efforts, which need to be applied to filling the military gaps left in the civilian research, not to duplicating that research. A better justification of the research approaches being pursued is needed in many areas of the intelligence research. A further justification in terms of military requirements is also needed.

Some of the approaches are becoming well developed enough for standardization—for example, mapping inside a building should now be standardized, and additional research should be aimed at object identification inside the building. ARL should lead a mapping effort in this area to ensure that all areas are covered and that relevant results are being leveraged. Incremental advances also include modeling the multiple-robot patrol problem in a new way and using machine learning in a variety of ways to improve robot intelligence.

ARL is currently employing simulation in its HRI research; this should be extended. There are additional roles for HRI beyond the testing of swarming robots. HRI should be considered in all aspects and stages of robotic research. Systematic tools for doing analysis of HSI needs should be used to drive definitions of mission and scenarios. ARL should take advantage of knowledge about human cognition in perception and intelligence applications. ARL should use more real robots and consider testing at Fort Benning, Georgia, with intended user groups.

Although the long-term vision is for soldiers to have robotic teammates, in the midterm, robots could be used as tools, with functions and tasks allocated according to supporting analyses of human and robot capabilities. The utilization of a robot as a trusted team member rather than as a tool is a noble goal that the ARL programs have embraced. HRI research should be given priority very early in all robotics programs.

ARL is to be complimented for the range of robotics sizes and the different types of mobility devices in its robot research portfolio. Also, the research portfolio for robotic mobility shows a good balance of analysis and physics-based modeling and experiments. It addresses real-world effects and has great focus on meeting specific needs. Metrics were fairly well defined, and the inherent requirement of a test vehicle drove the system thinking and approach. The work on the micro flyers and that on legged robotic systems are best in class. There was a good portfolio of vehicles and platforms from small scale to mesoscale and microscale. Certain efforts were judged to be leading the state of the art in their areas. Staff were aware of the system perspective and addressed it. One area that could be emphasized is more discrete awareness of mission, sensors, and power requirements to meet the application vision and scenario. However, the efficiency of existing robotic systems in transferring energy from the engine to the environment is still several orders of magnitude worse than that of biological devices, and therefore, continued work in this area is required. Because of the burden imposed on soldiers by battery packs and the limited time on mission for robots, research that improves the overall energy density and efficiency of converting energy to motion is required. In particular, research on the combustion of small JP-8 combustion engines and research on the efficient creation and transfer of force to the environment should be added to the research portfolio.

CROSSCUTTING ISSUES

ARL leadership is in transition. At several levels, from that of the ARL Director through individual directorates, "acting leadership" is the watchword of the day. The hard work and significant accomplishments of the current acting leaders are acknowledged, but instability introduces uncertainty, which in turn introduces the risk of inefficiency and misdirection. The Army should expedite a return to a state of stability of the technical management at ARL in the near future.

The hiring issue is not confined to senior management. ARL has been highly successful in recent years in recruiting many bright, early-career scientists and engineers, often in newly developing technical areas. These new recruits offer great promise for the future, but they are in need of strong technical leadership. Some technical areas are benefiting from seasoned internal leadership, but in a number of newer areas, senior technical leadership should be recruited from outside ARL, because ARL continually addresses emerging scientific and technical areas. Acknowledging limited flexibility in the Senior Scientific/Professional personnel track and issues of hiring freezes, ARL should consider appointments through the Intergovernmental Personnel Act process.

ARL has released the first volume of the "Research @ ARL" series. This event deserves congratulations. Focused on recent advances in energy and energetics, this volume of technical papers is the first of many planned documents that will be produced across the ARL directorates and that will help stakeholders understand the scope and direction of recent accomplishments by the dedicated and talented staff at ARL. Although addressing this audience through such technical compendia is desirable and praiseworthy, it should not be viewed as sufficient to capture the impacts of ARL's research. The full impact of research can only be measured after the fact. ARL can benefit from having a historian with sanctioned access as well as requirements for internal reporting, organized to ensure the collection of appropriate data and personal recollections.

The increased attention by ARL to enterprise R&D efforts is commendable. As the technical quality and depth within directorates continue to improve, it is appropriate that ARL continue to increase its focus on broad, multidisciplinary issues that can best be addressed by collaborative work across several directorates and with extramural partnerships that enhance the ARL intramural capability. It is becoming increasingly clear that greater attention should be given to the review of the work done by all participants in these collaborative alliances, both intra- and extramural. If the alliances succeed as intended, then their efforts have to make a profound impact on the content and quality of the ARL portfolio as well as on the accomplishments of its staff. If this is so, management should welcome the validation from external review. ARL should consider establishing an independent review that will allow adequate attention to the work done by all parties in the collaborative alliances. Additionally, it is difficult to find documentation that clarifies the advantages and disadvantages of these approaches to collaborative research and their comparative value to the traditional use of government laboratories by the Air Force and Navy. ARL should consider performing or commissioning retrospective analyses of these extramural collaborative activities, to be targeted at such issues as best practice in management, technical accomplishments, and impacts on the Army and on the conduct of business in the ARL.

1

Introduction

This introductory chapter describes the biennial assessment process conducted by the National Research Council's (NRC's) Army Research Laboratory Technical Assessment Board (ARLTAB). It then describes the preparation and organization of the report, the assessment criteria, and the approach taken during the report preparation.

THE BIENNIAL ASSESSMENT PROCESS

The charge of ARLTAB is to provide biennial assessments of the scientific and technical quality of the Army Research Laboratory (ARL). These assessments include the development of findings and recommendations related to the quality of ARL's research, development, and analysis programs. The ARLTAB is charged to review the work of ARL's six directorates but not the work of the Army Research Office (ARO), a key element of the ARL organization that manages and supports basic research; however, all ARLTAB panels receive reports of how the research and development activities of ARO and ARL are coordinated. At the discretion of the ARL Director, the ARLTAB reviews selected portions of the work conducted by the Collaborative Technology Alliances (CTAs). Although the ARLTAB's primary role is to provide peer assessment, it also may offer advice on related matters when requested to do so by the ARL Director; such advice focuses on technical rather than programmatic considerations. The ARLTAB is assisted by six NRC panels that focus on particular portions of the ARL program. The ARLTAB's assessments are commissioned by ARL itself rather than by one of its parent organizations.

For this assessment, the ARLTAB consisted of seven leading scientists and engineers whose collective experience spans the major topics within ARL's scope. Six panels, one for each of ARL's directorates,[1] report to the ARLTAB. Six of the ARLTAB members serve as panel chairs. The panels range in size from 14 to 22 members, whose expertise is carefully matched to the technical fields covered by the directorate(s) that they review. In total, 110 experts participated, without compensation, in the process that led to this report.

The NRC appoints the ARLTAB and panel members with an eye to assembling slates of experts without conflicts of interest and with balanced perspectives. The 110 experts include current and former executives and research staff from industrial research and development (R&D) laboratories, leading academic researchers, and staff from Department of Energy national laboratories and federally funded R&D centers. Twenty-seven of them are members of the National Academy of Engineering, and 4 are members of the National Academy of Sciences. A number have been leaders in relevant professional societies, and several are past members of organizations such as the Army Science Board and the Defense Science Board. The ARLTAB and its panels are supported by NRC staff, who interact with ARL on a continuing basis to ensure that the ARLTAB and panels receive the information that they need to carry out their assessments. ARLTAB and panel members serve for finite terms, generally 4 to 6 years, so that viewpoints are regularly refreshed and the expertise of the ARLTAB and panel members continues to match the ARL's activities.

Biographical information on the ARLTAB members appears in Appendix B, along with a list of each panel's members.

PREPARATION AND ORGANIZATION OF THIS REPORT

The current report is the seventh biennial report of ARLTAB. Its first biennial report was issued in 2000; annual reviews were issued in 1996, 1997, and 1998. As with the earlier reviews, this report contains the ARLTAB's judgments about the quality of ARL's work (Chapters 2 through 7 focus on the individual directorates, and Chapter 8 provides a discussion of crosscutting issues across all of ARL). The rest of this chapter explains the rich set of interactions that support those judgments.

The amount of information that is funneled to the ARLTAB, including the evaluations of the recognized experts who make up the ARLTAB's panels, provides a solid foundation for a thorough peer review. This review is based on a large amount of information received from ARL and on interactions between ARL staff and the ARLTAB and its panels. Most of the information exchange occurs during the annual meetings convened by the respective panels at the appropriate ARL sites. Both at scheduled meetings and in less formal interactions, ARL evinces a very healthy level of information exchange and acceptance of external comments. The assessment panels engaged in many constructive interactions with ARL staff during their annual site visits in 2011 and 2012. The introductory sections of Chapters 2 through 7 provide the dates of the panel site visits for each directorate. In addition, useful collegial exchanges took place between panel members and individual ARL investigators outside of scheduled meetings as ARL staff members sought additional clarification about panel comments or questions and drew on panel members' contacts and sources of information.

Each panel meeting lasted 2.5 days, during which time the panel members received a combination of overview briefings by ARL management and technical briefings by ARL staff. Prior to the meetings, panels received extensive materials for review, including selected staff publications.

[1]The six ARL directorates are the Computational and Information Sciences Directorate (CISD), Human Research and Engineering Directorate (HRED), Sensors and Electron Devices Directorate (SEDD), Survivability/Lethality Analysis Directorate (SLAD), Vehicle Technology Directorate (VTD), and Weapons and Materials Research Directorate (WMRD). Appendix A provides information summarizing the organization and resources of ARL and its directorates.

The overview briefings brought the panels up to date on ARL's long-range planning. This context-building step is needed because the panels are purposely composed mostly of people who—while experts in the technical fields covered by the directorates(s) that they review—are not engaged in work focused on ARL matters. Technical briefings for the panels focused on the R&D goals, strategies, methodologies, and results of selected projects at the laboratory. Briefings were targeted toward coverage of a representative sample of each directorate's work over the 2-year assessment cycle. Briefings included poster sessions that allowed direct interaction among the panelists and staff of projects that either were not covered in the briefings or had been covered in prior years.

Ample time during both overview and technical briefings was devoted to discussion, which enabled panel members to pose questions and ARL staff to provide additional technical and contextual information to clarify panel members' understanding. The panels also devoted sufficient time to closed-session deliberations, during which they developed findings and identified important questions or gaps in panel understanding. Those questions or gaps were discussed during follow-up sessions with ARL staff so that the panel was confident of the accuracy and completeness of its assessments. Panel members continued to refine their findings, conclusions, and recommendations during written exchanges and teleconferences among themselves after the meetings.

In addition to the insights that they gained from the panel meetings, ARLTAB members received exposure to ARL and its staff at ARLTAB meetings each winter. The 2011 ARLTAB meeting focused on the ARL crosscutting research areas, and the 2012 ARLTAB meeting focused on refining elements of the assessment process, including read-ahead materials, review agendas, and expertise required within the panels.

ASSESSMENT CRITERIA

During the assessment, the ARLTAB and its panels considered the following questions posed by the ARL Director:

1. Is the scientific quality of the research of technical quality comparable to that executed in leading federal, university, and/or industrial laboratories both nationally and internationally?
2. Does the research program reflect a broad understanding of the underlying science and research conducted elsewhere?
3. Does the research employ the appropriate laboratory equipment and/or numerical models?
4. Are the qualifications of the research team compatible with the research challenge?
5. Are the facilities and laboratory equipment state of the art?
6. Does the research reflect an understanding of the Army's requirement for the research or the analysis?
7. Are programs crafted to employ the appropriate mix of theory, computation, and experimentation?
8. Is the work appropriate to the ARL niche?
9. Are there especially promising projects that, with application of adequate resources, could produce outstanding results that could be transitioned ultimately to the field?

Within the general framework described above, the ARLTAB also developed and the panels selectively applied detailed assessment criteria organized in the following six categories (Appendix C presents the complete set of assessment criteria):

1. *Effectiveness of interaction with the scientific and technical community*—criteria in this category relate to cognizance of and contributions to the scientific and technical community whose activities are relevant to the work performed at ARL;
2. *Impact on customers*—criteria in this category relate to cognizance of and contributions in response to the needs of the Army customers who fund and benefit from ARL R&D;
3. *Formulation of projects' goals and plans*—criteria in this category relate to the extent to which projects address ARL strategic technical goals and are planned effectively to achieve stated objectives;
4. *R&D methodology*—criteria in this category address the appropriateness of the hypotheses that drive the research, of the tools and methods applied to the collection and analysis of data, and of the judgments about future directions of the research;
5. *Capabilities and resources*—criteria in this category relate to whether current and projected equipment, facilities, and human resources are appropriate to achieve success of the projects; and
6. *Responsiveness to the ARLTAB's recommendations*—with respect to this criterion, the ARLTAB does not consider itself to be an oversight committee. The ARLTAB has consistently found ARL to be extremely responsive to its advice, so the criterion of responsiveness encourages discussion of the variables and contextual factors that affect ARL's implementation of responses to recommendations rather than an accounting of responses to the ARLTAB's recommendations.

APPROACH TAKEN DURING REPORT PREPARATION

This report represents the ARLTAB's consensus findings and recommendations, developed through deliberations that included consideration of the notes prepared by the panel members summarizing their assessments. The ARLTAB's aim with this report is to provide guidance to the ARL Director that will help ARL sustain its process of continuous improvement. To that end, the ARLTAB examined its extensive and detailed notes from the many ARLTAB, panel, and individual interactions with ARL during the 2011-2012 period. From those notes it distilled a shorter list of the main trends, opportunities, and challenges that merit attention at the level of the ARL Director and his management team. The ARLTAB used that list as the basis for this report. Specific ARL projects are used to illustrate these points in the following chapters when it is helpful to do so, but the ARLTAB did not aim to present the Director with a detailed account of 2 years' worth of interactions with bench scientists. The draft of this report was subsequently honed and reviewed according to NRC procedures before being released.

The approach to the assessment by the ARLTAB and its panels relied on the experience, technical knowledge, and expertise of its members, whose backgrounds were carefully matched to the technical areas within which the ARL activities are conducted. The ARLTAB and its panels reviewed selected examples of the standards and measurements activities and the technological research presented by ARL; it was not possible to review all ARL programs and projects exhaustively. The ARLTAB's goal was to identify and report salient examples of accomplishments and opportunities for further improvement with respect to the technical merit of the ARL work, its perceived relevance to ARL's definition of its mission, and apparent specific elements of ARL's resource infrastructure that are intended to support the technical work. Collectively, these highlighted examples for each ARL directorate are intended to portray an overall impression of the laboratory while preserving useful mention of suggestions specific to projects and programs that the ARLTAB considered to be of special note within the set of those examined. The ARLTAB applied a largely qualitative rather than quantitative approach to the assessment. The assessment panels' site visits are currently scheduled to be repeated annually, and the assessment report to be issued biennially.

REPORT CONTENT

This chapter discusses the biennial assessment process used by ARLTAB and its six panels. Chapters 2 through 7 provide detailed assessments of each of the six ARL directorates. Chapter 8 presents an overview focused on crosscutting issues across all of ARL. The appendixes provide the ARL organizational chart, biographical information on the ARLTAB members, a list of the panel membership, the assessment criteria used by ARLTAB and its panels, and a list of acronyms found in the report.

2

Computational and Information Sciences Directorate and Network Science Enterprise

INTRODUCTION

This report is based on meetings of the Panel on Digitization and Communications Science held at ARL on June 28-30, 2011, and June 6-8, 2012. The 2012 meeting was expanded to include a review of the network science portfolio of activities on which the Computational and Information Sciences Directorate (CISD), Human Research and Engineering Directorate (HRED), and the Sensors and Electron Devices Directorate (SEDD) collaborate. The panel reviewing the network science activities included members selected from the panels that also reviewed SEDD and HRED as well as members of the panel that reviewed CISD.

This chapter includes the report on the Network Science Enterprise. Two divisions in CISD, Information Sciences and Network Sciences, have substantial involvement in the Network Science Enterprise, and the reports on those two divisions are subsumed under that for the Network Science Enterprise. Separate sections are devoted to the Battlefield Environment Division and to the Computational Sciences Division, the remaining two divisions of CISD, which have considerably less involvement in the Network Science Enterprise. Discussions and recommendations on opportunities and challenges, and on the overall technical quality of the work, are presented in the respective sections, including the network science section. The chapter ends with common themes of opportunities and challenges, and observations on the overall quality of the work that cut across the Network Science Enterprise and CISD.

Two major extramural collaborations involve both the Network Science Enterprise and CISD: the International Technology Alliance (ITA), which is an alliance of the United Kingdom (UK) Ministry of Defence, universities, and industry in the UK and United States; and the Network Science Collaborative Technology Alliance (CTA), which has participation from CISD, HRED, and SEDD in ARL as well as a number of universities and industry partners. The ITA and the CTA are representative of "enterprise" collaborations within ARL—a concept that is, quite appropriately and laudably, receiving increasing attention and momentum within ARL. The ITA was initiated in 2008 and the CTA in 2010; this CTA follows a previous CTA, which also had a network focus.

COMPUTATIONAL AND INFORMATION SCIENCES DIRECTORATE

Changes Since the Previous Review

During this period, a new director took the helm of CISD, and a new chief took the helm of the Computational Sciences Division. These leadership changes have benefited the respective organizations by injecting continuity and stability and have led to sharper focus of the research vision and activities. Particularly noteworthy in CISD is the strength of the aspiration to increase activities across disciplines, divisions, and other organizations. Also, in the Computational Sciences Division (CSD) there have been definite improvements in direction and focus that may be reasonably credited to these management changes.

ARL should take note of the potentially destabilizing and morale-lowering effects of overly rapid turnover of management and of having acting appointees head organizations for extended periods of time. Minimizing such disruptions will greatly benefit ARL.

Battlefield Environment Division

Introduction

The Battlefield Environment Division (BED) presented a compelling picture of the need for basic research in atmospheric science to meet Army needs, which range from aiding special operations to improving the accuracy of conventional artillery. The Army has a unique and pressing requirement for near-Earth atmospheric understanding and characterization beyond what can be provided by other military and civilian entities. The presence of 24 Ph.D.'s out of a total of 56 civilian divisional staff and the ongoing effort to increase staff expertise through the involvement of postdoctoral personnel are impressive developments. The division's support appears to be a reasonably good mix of 6.1, 6.2, and customer funding, with the first two constituting 55 percent and the last 45 percent. Seventeen percent of customer funding is from the Defense Advanced Research Projects Agency (DARPA). Total program support dropped by approximately 10 percent because of cutbacks in the area of atmospheric sciences applications, which is not surprising given the advances that have been accomplished in previous years.

Collaborations in the Mountain Terrain Atmospheric Modeling and Observations (MATERHORN) Program and Partners In Research Transition (PIRT), a joint program with various organizations and universities, illustrate a vigorous extramural activity. There were a number of personnel initiatives during the past year, including establishing a relationship with a professor at Colorado State University and rotating a BED staff member to the Army Research Office to help in the atmospheric science area.

BED's research focuses on three major areas: atmospheric sensing, atmospheric dynamics, and atmospheric modeling applications. Underlying these activities is fundamental science, with the emphasis currently on four areas: near-field phase locking of array laser, Raman spectra of individual particles, turbulence propagation, and atmospheric impact routing. A significant portion of the division's resources are directed at products that it will provide to the Army. Management has a technical roadmap for the division that extends to 2017 and that deals with major programs currently in-house and others that are anticipated. Overall, the division's management provided compelling evidence of a grasp of the challenges, especially the balancing of diverse requirements with limited and shrinking resources, and the planning and vision to address the challenges.

Accomplishments and Advancements

Raman Spectra of Individual Airborne Particles Several optics projects in BED are aimed at single-particle detection for biohazard threat applications. BED has made significant contributions in the area of laser-induced fluorescence (LIF) detection methods that have been developed and now deployed, but these LIF methods do not provide adequate differentiation to reliably identify specific pathogens. Definitive characterization is more straightforward for chemical detection using mass spectrometry techniques, for example, but pathogen-specific biosensors in atmospheric sensing require techniques such as Raman spectroscopy or infrared (IR) absorption/emission spectroscopy. Both of these techniques suffer from signal-to-noise constraints at the single-particle level, and BED is engaged in basic early-stage research aimed at addressing this challenge. For Raman spectroscopy, current work provides a novel method for trapping individual particles using a new cylindrical (Bessel) beam-focusing geometry that provides a hollow optical ellipsoidal intensity chamber containing particles through photophoretic forces. This provides sufficient residence time (approximately 1 second) yielding the signal-to-noise ratio required for reliable pathogen identification. This technique has already demonstrated the ability to trap a significant range of particle sizes using readily available milliwatt-level beam powers along with practical optics, and it has also demonstrated well-resolved Raman signatures of trapped particles using the same trapping beam. This technique also has the flexibility of allowing for additional specific Raman probe beams. The work is in the very early stages, but it has demonstrated that in airborne particle applications the photophoretic forces are much stronger than the usual optical tweezer radiation pressure techniques that are used routinely in aqueous environments. There is opportunity for obtaining a wealth of data and for data analysis, combined with theoretical modeling, to fully map out the efficacy of the approach. Once this is known, the group has competence in the hardware implementations necessary to demonstrate a combined system that uses preliminary scattering and LIF-based particle sorting, followed by the Raman technique for specific pathogen identification.

Turbulence Propagation Theory and Effects This study has matured very nicely. The principal investigators have published results and have used reviewer feedback to develop a more complete theory for the modulation transfer function (MTF). The present study has two objectives: (1) quantification of optical turbulence effects in passive and active imagers and (2) development of prototype methods for mitigating turbulence effects on imagers. With regard to the first objective, the research team should proceed with publication of the new (more complete) theoretical findings. With regard to the second objective, the researchers have begun to develop concepts for operationalizing the theoretical results. The research team should develop a prototype system to demonstrate that the theory correctly describes what happens in practice, and it should use the results to design and implement a full demonstration program to present to a funding agency within the Army. The most recent research is unique and fundamental; unlike the adaptive optics approach also developed within BED, this approach does not require a cooperative target to enhance the target image.

Climatological Assessment Utilizing Airborne Acoustic Sensors The problem addressed in this activity is locating a source of acoustic emission on Earth's surface using detectors on an elevated/airborne platform. Once the location of the emission is identified, a camera can be directed toward the source for full identification. The path of the propagating acoustic signal has to be corrected for refraction or bending resulting from temperature variations in the atmospheric density and shifts due to horizontal wind. The researchers lucidly defined the problems, and they tested the corrections for a case of heated ground that has a monotonically changing temperature profile concave with respect to Earth's surface.

This study involves one particular case that may be representative of some situations occurring in the Afghan theater but not to all possible situations, such as those with very stable conditions or an inversion between the Aerostat and the ground, which could happen, especially when the Aerostat is positioned at 1,600 m (cases were presented for 200, 800, and 160 m) altitude. The approach developed a climatology (based on 155 days of data) of near-surface wind and temperature profiles and extracted the needed corrections using the corrections based on published theoretical calculations. The method was tested using data from two field sites, Aberdeen and Yuma Proving Grounds. The data revealed a smooth temperature curve with height but did not show any inversion, a result that might make the method invalid for some situations and may suggest a need for further testing. The developments appear reasonable, perhaps because the requirement for a camera to aim in approximately the correct direction is not very severe, so that even though these early tests show considerable scatter in the corrective terms, the results point in the right direction.

Tests using forecast profiles are planned, which is an important step to establishing and applying the method operationally. Attention should still be given to situations with a stable air-layer near the ground and a possible inversion between the Aerostat and the ground. Further studies on the adequacy of climatologically stored data or local forecasts for estimating the corrections of the localization angle and distance are needed and should answer the question regarding the method's applicability in the field.

Weather Research and Forecast Model-Based Nowcasting for Battlefield Operations BED's weather research and forecast (WRF)-based numerical modeling is designed to fill an important gap that exists at very fine scales, in order to meet Army needs for timely environmental data on the rapidly evolving battlefield. This need is not currently met well by standard, available weather forecast products designed primarily for larger spatial scales and longer temporal scales. BED's numerical "nowcasting" effort is designed to provide frequent updates of environmental products on grids of 1 kilometer or less. To be successful, this work has to assimilate, in addition to traditional sources of weather data, a variety of atmospheric observations collected by disparate sensors on the battlefield, and it has to be able to identify and reject unreliable data. BED researchers are investigating ways to exploit satellite data that could provide models with soil-moisture measurements, which would be valuable for improving the accuracy of land-surface flux calculations and atmospheric boundary layer conditions. Traditionally, these high-resolution modeling tasks have required major high-performance computing platforms, but BED is examining means of achieving forward-deployed computing capacity.

Currently, the means to accurately represent the effects of turbulence in the atmospheric boundary layer at the scales relevant for high-resolution nowcasting is not well understood and is poorly represented by available modeling software. BED has made progress in recognizing this problem and in developing and/or acquiring the necessary expertise to address this issue. A hierarchy of models is under development and is being tested for application in various battlefield conditions. This logical path allows for steady improvement of atmospheric guidance as advances are made, the science matures, and computational capacity increases.

BED has made significant strides in addressing the issues unique to weather model verification at the battlefield scale by acquiring and evaluating advanced weather model evaluation software. This effort is already stimulating new ways to extract and evaluate the accuracy of Army-relevant information from the guidance provided by the weather model. BED's modeling work supports a wide range of products that assist the soldier, including optimal routing algorithms that address impacts of atmospheric conditions, improved accuracy of ballistic artillery, and soldier health effects such as heat stress and exposure to airborne toxic aerosols.

Although the numerical modeling effort is making substantial progress toward solving several important problems, it faces a few significant hurdles. WRF applications at sub-kilometer grid scales, especially in unstable conditions, are problematic, because of limitations to the available turbulence physics parameterizations. Data assimilation at very fine scales is handicapped by the paucity of observations at comparable scales. BED is encouraged to pursue this legitimate but challenging research area. Many of the modeling initiatives undertaken during the past year have involved developing new techniques as well as importing of new expertise. It is not surprising, then, that BED has not yet had sufficient opportunity to perform adequate model verification, which is beginning to emerge from these recent developments, and which BED intends to continue.

Atmospheric Impacts Routing Tool This project is a continuation of an effort that has been presented in two prior reviews. This tool has matured and is ready to move beyond the prototype stage into operations. It applies the "A*" algorithm (a computer algorithm that is used in path finding and graph traversal) to the problem of "optimal" air and ground transportation routing. Ground routing involves multiple impact factors, such as Met forecasts, nowcast information, vehicle types, soil conditions, road states, and other factors. The algorithm "optimizes" paths in three-dimensional (3D) or four-dimensional (4D) space after building an architecture to support the ingestion of dynamic 3D data sets. It then makes an informed search for the "best" route, applying a set of criteria for what is best. It can adjust for desired goals, such as wanting higher than normal safety or faster speed to the goal point. Areas of avoidance can be declared as "no-go"/"no-fly" areas.

The project has been extended to respond not only to weather conditions but also to other threats or obstacles that could impede or preclude use of a particular route. The technology appears to run very fast, calculating optimized (for example, minimal time) routes while avoiding adverse conditions. It avoids many of the problems associated with pre-defined networks and instead finds a solution for an optimal path through a 3D grid, subject to the constraints imposed by the user or the different threats; it was originally based upon the A* algorithm but has been expanded to incorporate the D* and E* algorithms.

Of concern is that only a few methods can verify that the routing system is performing correctly, that is, generating an optimal route given a particular 3D arrangement of threats. One possible solution involves configuring the system to generate all possible routes through a given 3D grid, and then determining how the ensemble varies statistically from the optimal one. Although the results of the A* algorithm are very impressive, the development of a method for testing optimal routing techniques might represent an even more significant improvement. Also, this research has the potential to benefit from cross-disciplinary expertise, especially in optimization and computational methods for large-scale optimization, which is likely to exist in other divisions, such as CSD. Additionally, it would be beneficial if the research considered the value of clearly defined "near-best" and "near-optimal" solutions.

Optical Systems A project in early-stage development involves coherent, phased-array beam combining using single-seed laser sources and arrays of fiber amplifiers with antenna elements controlled both by individual phase modulators and piezo-actuated fiber positioners. This Intelligent Optics Laboratory (IOL) system has already demonstrated record power-combining at the multi-kilowatt level, including active servo control of the transmitter for some degree of atmospheric turbulence correction using element phase and beam tilt adjustment, with elaborate optimization of servo control to counter transit-time-induced loop delays. The team's several hardware approaches demonstrated scalability, both in element power and number of elements, which has the potential to take the solution to weapons-level powers. The significant customer pull here (DARPA) provides strong evidence of the leading-edge nature of this work.

Adaptive Optics The IOL team also was involved in imaging through boundary layer turbulence using locally engineered adaptive optics hardware that does not rely on wavefront sensing and has potential to constitute a low-cost solution. Using a novel, laterally actuated, low-piezo element count, 1-inch diameter adaptive optics, the IOL team has demonstrated performance that is close to that of much more elaborate, and very expensive, commercial solutions. The team also has explored the feedback servo problem, including removing loop delay. Work in this area is highly promising, and further exploration will yield valuable assessment of different feedback signals, such as image spatial frequency content, that provide best image fidelity in a highly dynamic environment. This work provides a good example of how BED is identifying special Army needs, in this case stemming from boundary layer turbulence, that require hardware solutions that cannot simply be adopted from existing astronomy or high-altitude surveillance solutions.

These turbulent imaging and beam transmission experiments, as well as the theoretical studies of image turbulence and hardware solutions (see the section "Turbulence Propagation Theory and Effects," above), are extremely important to fulfilling BED's mission. Because the problem of imaging and transmission through boundary layer turbulence is somewhat unique to the Army's mission, it is not clear that adequate resources and attention are being applied across the DoD to solving this problem in order to deliver the advantage that may be possible using constantly improving hardware and low-power graphical processing hardware. This makes BED's works even more critical.

Applied Anomaly Detection The work on applied anomaly detection is part of the atmospheric science initiative within BED, and it specifically links the social-cognitive and information genres. It has both 6.2 funding and customer-driven funding sources.

This project is noteworthy because it brings together cognitive processing experts, ARL machine learning experts, sensor experts, and a military expert with the goal of giving the soldier the capability and training to sense a dangerous situation in a field. The objective is to transfer to a machine the military expert's ability to find anomalies that signal possible danger in a complex or cluttered field. The team has developed techniques that provide an approach for training the machine and thereafter soldiers to perform at the expert level. Image features such as shape, color, contrast, and texture are used to discriminate anomalies. The Applied Anomaly Detection Tool software incorporates the team's techniques and is the main enabler of dissemination. The tool is designed for training soldiers through computer-based detection exercises.

This solid work occupies an important niche that is matched to the Army's needs, and it blends the strengths of several disciplines. The research should reflect greater awareness and absorption of learning from related work around the world. Given its potential, ARL should encourage this activity and extend it to other applications that can benefit from capturing human, especially soldiers', expertise.

Opportunities and Challenges

Artillery Meteorology The Army and Marine Corps artillery communities are finally actively pursuing significant upgrades to the BED-supported system for artillery meteorology. The next-generation system, Computer Meteorological Data Profiler (CMD-P), which BED claims to be revolutionary, is built around a yet-to-be-tested dynamic microscale model derived from the widely used WRF modeling system. The ability to assimilate a variety of nontraditional observations (e.g., opportunity-met observations from a passing unmanned aerial vehicle) is a key element of the modeling system. The described work is in an early stage. Success in this endeavor will require addressing several challenges, particularly in light of the amount of computing power likely to be available in the field. Using WRF at the desired 0.5 km spatial resolution will push against the limits of applicability of the physics (e.g., turbulence closure scheme) in the model. Model development should proceed in parallel with a verification process, and the model should be designed to be easily upgraded as advances are made in the WRF model by the wider meteorological community. Some aspects of this work parallel efforts to adapt the WRF to modeling small-scale effects in the urban boundary layer; the CMD-P2 developers should carefully monitor these parallel efforts. BED developed, maintains, and supports a number of numerical mesoscale and microscale models. It would be advantageous for BED to develop a common modeling system/framework rather than a series of unique models.

Verification and Validation of Models Consistent, appropriate verification and validation of mesoscale and microscale models remains a challenge for the wider meteorological numerical modeling community. BED is still at an early stage in arriving at an assessment scheme for its various numerical prediction models. The exploration of the NOAA/NWS RTMA product[1] is a reasonable, but only a first, step. BED should reconsider the NCAR DTC MET system[2] and explore some of the other assessment tools that have been developed in university modeling programs. This effort is a vital one for the BED modeling team. The credibility of and trust in the performance of BED-developed models needs to be based on a solid, well-accepted verification and validation scheme.

Need for New Physics in the Modeling System BED's work to apply the WRF model to small spatial scales (for example, 1.0 to 0.5 km) appears to be closely related to that of Artillery Meteorology and other soldier-scale environmental prediction projects (see the section "Weather Research and Forecast Model-Based Nowcasting for Battlefield Operations," above). Most importantly, this effort is pushing up against the physics of the modeling system, which represents an opportunity for BED to make a fundamental contribution to numerical modeling at the meso- and microscales, given its expertise in the area of empirical observations of turbulence in small regions. A new, well-verified turbulence closure scheme appropriate for small scales for use in the WRF model would be welcomed by the wider modeling community. Again, this development effort should be supported by a parallel verification effort. The NOAA Meteorological Assimilation Data Ingest System (MADIS) system has latencies that are too long to use with a meso-/microscale modeling system supporting nowcasting.

Anisotropic Turbulence Anisotropy in turbulence near Earth's surface represents a fundamental problem of boundary layer meteorology that requires characterization for updating our very simplistic isotropic view of turbulence theory. The case of strong anisotropy needs to be parameterized for practical use. Turbulent features in complex terrain and urban environments can be very important for transmission path calculations, unmanned aerial vehicle operations, and wind energy planning. Continuing work toward making this type of information accessible and easy to understand and use is encouraged. A good step in this direction would be to reach outside the community by publishing a fully peer-reviewed paper on this subject. The work is very interesting and has practical application for the Army in the field/theater and on its bases, as well as for its use of wind turbines for some of its energy needs. The knowledge would also have valuable civilian applications. BED mentioned planned use of lidar to measure turbulence structure in the atmosphere. Measurements at elevations on the NOAA Boulder Atmospheric

[1]Real-Time Mesoscale Analysis (RTMA) products from the National Oceanic and Atmospheric Administration's (NOAA) National Weather Service (NWS).

[2]NCAR DTC MET system stands for the Model Evaluation Tools (MET) verification package developed by the National Center for Atmospheric Research (NCAR) Developmental Testbed Center (DTC).

Observatory tower, higher than provided in the earlier data sets, and the promised tethered turbulence sensors, should allow further characterization in terms of height dependence of these anisotropic structures. Another interesting avenue of continued work is the study of Fourier transformation use for these turbulent temporal or spatial data series (even though the data are not derived from wave forms), which could have very interesting theoretical ramifications in time.

BED might be approaching the time when its high-quality work in adaptive optics can provide atmospheric parameters to the atmospheric modelers and experimenters to improve their ability to generate spatially dense estimates of the near-Earth atmosphere state. BED should direct attention to this possible synthesis of information.

As BED refines its strategy to handle opportunities and challenges, it should expand its good working relationships with university groups to include those that are strong in fundamental modeling research. The modeling challenges, especially the ones mentioned above, are substantial, and extending the university ecosystem to include partners that are strong in fundamental modeling research will be valuable. BED should also welcome industry partnerships. BED should consider the possibility of working with ARO to establish a center of excellence that would bring together appropriate partners from universities and industry, to contribute to the effort in which it is both a leader and a player with a large stake.

Overall Technical Quality of the Work

BED's management and staff members are to be congratulated for their progress toward becoming a first-class research organization, especially in times when the kind of research they are conducting is under stress. The presentations delivered, both orally and through posters, were of very high quality and represented a significant concentration of effort on solving fundamental scientific problems. Even the warfighter-focused applications contained a high degree of engineering research leading to noteworthy achievements. Discussions with division personnel revealed significant pride in the division and in its research, as well as very good morale, which are essential to continued success.

Computational Sciences Division

Changes Since the Previous Review

High-performance computing (HPC) is an essential enabling capability that supports many ARL research programs. Many ARL projects presented during the 2011 and 2012 reviews, including those from outside the CSD, involve very large data sets or require extensive computation. CSD's HPC resources and its ongoing computational research are important to ARL's research more broadly. HPC's role in various ARL research projects and Army applications has grown since the 2009-2010 review, and it is expected to continue to grow. Emerging and future advances in computational science will expand the applicability and utility of HPC to ARL and other Army research.

Many of the ARL projects involve models of some form. In the 2009-2010 reviews, model verification and validation were often absent or insufficient. In the 2011-2012 reviews, many of CSD's projects provided evidence of increased attention to the importance of verification and validation, and to the appropriate methods of performing these. In general, the overall technical quality and focus of CSD's nascent research program is improving rapidly as it gains maturity and coherence.

Accomplishments and Advancements

CSD's structure includes both a substantial facilities component that serves the HPC and networking infrastructure needs of ARL, the Army, and the DoD, and a new and growing research component focused on interdisciplinary computational science. The CSD team has made substantial progress in a very short time in articulating a research vision and realigning activities to support that vision within the research component.

CSD's research component has three major thrusts. Simulation science leverages existing CSD and ARL strengths along strategic areas of opportunity and impact including multiscale modeling (aligned with the computational materials science initiative), simulations of networks (aligned with the network science initiative), and biological HPC. Computational architectures is aimed at the development of computing architectures that support Army missions, including tactical cloudlets, hybrid computing, and power aware computing. Computer science and high-performance networking encompasses computational environments, HPC solutions for large data set analytics, leveraging the ARL Enterprise Optical Network (AEON) for networking research, and new networking architectures. In general, CSD is appropriately focused on determining how to effectively deploy and exploit HPC capabilities, rather than on how to develop new HPC and networking hardware.

CSD's multiscale materials modeling program focuses on fundamental materials modeling research that requires intensive computation capabilities. In general, this program is seen as potentially very important because of its breadth of relevance to the Army mission, for example, from electronic and photonic materials and devices to new reactive armors. Care should be taken to preserve this essential connection to Army applications in this program's individual research projects. The combination of computational resources with important materials problems plays to the strengths of ARL and CSD. This program provides an excellent opportunity to identify and drive research in simulation science, especially in algorithmic innovations needed for scalability while meeting accuracy and stability requirements imposed by complex physics-based models. There was evidence that CSD understands that verification and validation is required not only for the models, but also for the data exchanges between the models and transitions between scales. However, this is a huge field, and focus is necessary to ensure that identifiable progress can be made in some selected areas with the available resources.

As this program moves forward, several important issues concerning the basic premise of multiscale modeling itself should be considered; two are mentioned here. The first issue pertains to the semantics of interconnecting models at different scales and the likely challenges that will arise in doing so. Passing output values from one model to the next model to use as input is not the real challenge. Rather, the challenge is to validate that the domains of applicability and assumptions of the interconnected models are compatible over their entire range of parameterization, and that they remain so even as those models are modified and adapted to new purposes. Attempting to do this in general for all applicable models is far too broad an objective for this program and should be avoided; instead, a real and feasible contribution would be to define the information-passing taxonomy. The second issue relates to the belief that modeling at the electron-atom level can in principle be extended to the continuum level by assuming that one can successfully bridge gaps between modeling scales. This presupposes, however, that the physics of the phenomena being modeled are known at all scales, which may not be true. For example, the atomistic-level calculations are based on energy minimization schemes, i.e., essentially the assumption of thermodynamic equilibrium. However, at the mesoscale, most materials systems are far from thermodynamic equilibrium and could be described using macroscopic non-equilibrium thermodynamics (which could be linear non-equilibrium thermodynamics, nonlinear thermodynamics, and extended thermodynamics).

One project within the multiscale materials modeling program explored the thermal performance of materials most relevant to the Army (such as the explosive RDX). Its novel simulation approach uses phonon-Boltzmann transport to bridge from the atomic (nano) to the continuum (meso) scale. Initial simulations results reported for 1-dimensional (1D) quartz were validated against experimental measurements of the specific heat of RDX. The initial 1D results were promising, with better agreement (a 7 percent discrepancy) than that produced by a full molecular dynamics model. A journal publication based on this work is still in preparation; only conference papers have been published thus far.

CSD's simulation of defects and dislocation dynamics project, which is part of the simulation science research thrust, focuses on computational simulation of defects and dislocations to enable the design of electronic materials of Army relevance. The project leverages an existing code for dislocation dynamics and couples it to an in-house finite element modeling code to provide new capabilities, such as surface detection and tracking of defects in gallium nitride (GaN) films. Model validation consists of comparing computational and analytic results, which is a good initial step. Further validation by comparing experimental data from observations with actual material is essential to fully establish model credibility and utility. The research results should be targeted for publications in top venues. An evaluation of alternatives to the project's distributed shared memory (DSM) approach for code coupling should be conducted; DSM could become a bottleneck for the scaling of simulations to model realistic time and length scales.

Explosions under vehicle bodies are currently a significant cause of casualties during Army operations. The objective of CSD's reduced order models for underbody blasts project is to execute blast predictions faster and cheaper, ideally on a cell phone. In general, high-fidelity models are expensive, are highly nonlinear, and have too many parameters for efficient calibration and optimization. This project's premise is that in many cases a significant portion of the parameter space belongs to a subspace with significantly lower dimension, i.e., fewer relevant parameters. The project uses a high-fidelity model and a design of experiments method to select the most important parameters and then Galerkin projection to reduce model dimension. The preliminary results seem promising, showing a sample reduction of degrees of freedom from 1.4 million to 50. These results were mainly reported in technical reports and presentations. The effectiveness of the method should be evaluated in comparison with that of other dimension reduction approaches when further results are produced.

Opportunities and Challenges

CSD's scientific research productivity and progress are substantial, given the relatively small number of personnel (21 full-time equivalents) devoted to research. However, the results of some projects oriented toward fundamental research have not been published in high-quality peer-reviewed venues. Furthermore, no reliable metrics are in place within CSD for evaluating the effectiveness or impact of software artifacts developed as part of, or as the primary goal of, CSD research projects.

Management should be sensitive to the possibility of competition within CSD for resources supporting HPC infrastructure versus performing basic research in computational science. The effort expended by CSD on general purpose software development should be measured and recognized. The decisions on resource allocation will have far-reaching impact. CISD should further invest in CSD with the intention of increasing the number of personnel dedicated to computational science research. However, recruiting may be difficult because of the many opportunities for computational scientists in other areas.

Not all CSD projects were well justified. All projects should be evaluated to ensure that the benefit, which should be specifically defined and quantified whenever possible, justify the cost (time and money). Furthermore, a project's context is important—What impact will it have? Where is it going?—and should be known and articulated.

Awareness of prior extramural work continues to be uneven. Some project presentations demonstrated that the researchers were very familiar with relevant previous research and software; unfortunately, others did not. In general, repeating work is a suboptimal use of scarce resources.

The division's real-time radio frequency propagation path loss project focuses on performing general purpose computing on graphics processing units to model path loss in radio frequency propagation, with a goal of modeling 5,000 devices in real time. The project uses Longley-Rice, transmission line matrix (finite difference), and ray tracing modeling methods. Considerable effort has been expended to transfer the FORTRAN code to another language. The models do not account for foliage, which would be very important at higher frequencies (for example, >500 MHz), and there is no incorporation of the signal bandwidth, which may be significant. The existing terrain topography databases may not be adequate for the calculations. The potential benefit of this research does not appear to have been quantified, and sufficient awareness of highly related prior art, for example previous brigade-level modeling in the Force XXI Battle Command Brigade and Below (FBCB2) project, was not evident. However, if the research is successful, then the models could be good tools for understanding system limitations.

CSD's tactical cloudlets program is based on a promising concept with the potential to deliver HPC "on demand" to the soldier, which could have a transformative impact on the Army's mission. However, there are major research challenges related to the designs of such a system and of a research program to support it. To better support the design of an extensible cloudlets system that can bring new capabilities to the Army of the future, CSD is encouraged to develop a research roadmap that includes both near-term (1-year) and long-term (5-10 years) goals, strategies, and milestones. The roadmap should include a list of specific planned or envisioned research tasks that both fit within the program's concept and support the Army's mission.

As this work continues and new short-term projects are selected along this theme, CSD should emphasize greater selectivity and prioritization of projects, with a focus on projects that can (in the near term) both deliver benefits to the soldier and provide insights into the design space of a broad tactical cloudlet effort. Finally, this effort does not fully conform to the emerging consensus definitions within the larger technical community for the terms "cloud" and "cloudlet." As currently formulated, this effort might be better described as "battlefield HPC," rather than as "tactical cloudlet." Consistency should be sought in describing the effort and in setting its research agenda. Clouds or caches will be deployed all over the battle areas. The big difference from earlier systems is that cloud systems are substantially larger in terms of storage, processing, and the ability to communicate with them, and clients operate much more autonomously. Basically, ARL needs a facility that is operated as a cloud with all that that implies. It would not be an HPC server, although the claim will undoubtedly be that such a private cloud exists.

As an initial seed project in a new area for ARL, CSD's relationship between brain structure and function project has a well-defined and reasonable scope and has made good progress. Its results show that multiple instances of ordinary differential equations previously used separately to simulate brain activity can influence each other in ways that qualitatively resemble linkages between brain regions. However, the presentation's descriptions of the project results as "validation," and the assertion that these results show that the model output "can be used as simulated brain signals," substantially overstate the project's current status. Validation by comparison to empirical data is required before those claims can be made. Every practicing modeler knows from experience, and it has been formally proven, that compositions of separately valid models are not automatically valid and need to be validated as a composition. Furthermore, the project's scale may be too small to initiate a new CSD research thread in traumatic brain injury, and if that is the goal, then a larger-scale effort may be needed. Collaboration with neuroscience or medical researchers is recommended for access to validation data and modeling guidance.

Overall Technical Quality of the Work

The overall technical quality of CSD's research is difficult to characterize at the division level, because it varies considerably from project to project. Moreover, time constraints prevented highly detailed reviews of the projects. Nevertheless, it can be safely said that the overall technical quality of CSD's research is clearly improving over time.

Some CSD projects and programs are of very high quality. For example, the multiscale materials modeling demonstrated a very good basic science approach—among the best seen during the reviews—and has ambitious and powerful goals that build on ARL strengths.

To the extent possible given security classification issues, ARL in general and CSD in particular should employ the mechanisms for confirming technical quality that are available in academic circles, for example, peer-reviewed publication, winning conferences' best paper awards, occupying conference leadership positions, such as the chairmanship of the technical program committee, and being the organizer of workshops that are associated with the conferences. Many indicators exist to properly assess research productivity and impact. The latter could be assessed, in part, on the basis of such quantitative metrics as the *h*-index of the more senior researcher, and adoption of software or recognition of leadership in professional societies.

NETWORK SCIENCE ENTERPRISE

Introduction

This section discusses the activities of the network science program in ARL and the Information Sciences Division and Network Sciences Division of CISD.

The Network Science Enterprise is beginning to take shape. It is very much a work-in-progress, but the changes that have occurred have been promising and directionally correct. Of particular note are the soaring aspirations of the "enterprise" concept, which is based on multiple network types, called genres, which span a very broad spectrum of activities that have networks in common, and linkages and dependencies across genres and directorates. Realization of the goals for network science will require close and active collaborations across the CISD, HRED, and SEDD directorates and a supporting management structure.

Table 2.1 lists the constituent genres, respective activities, and the partnering directorates in each genre.

The vision of abstracting common concepts and mathematical structures across genres is laudable. Aiming to exploit the potential of composite networks is worthy, timely, and ambitious. Underlying this

TABLE 2.1 Genres, Respective Activities, and Partnering Directorates in Each Genre

Genres	Activities	Partners
Social-Cognitive	Human cognition and decision making in networks; trust in distributed decision making	HRED and CISD
Information	Text, image, and video processing; quality-of-information aware networking; trust management; cyber-defense	SEDD and CISD
Communications, including its physical layer	Constrained environments; dynamic networks	CISD

vision is the understanding that each genre has its own realization, some parts of which already exist and others being researched; these genres have been conceived independently but depend on coordination with networks of other genres for operational use. In this setting it is reasonable to hypothesize that by taking a unified view and combining the goals and designs of two or more genres, superior performance and efficiencies may follow, especially if the execution is initiated early and close to the conceptual stage and then followed through. This is the vision of the Network Science Enterprise.

The challenges of such combination are large. The difficulties of working across cultures, disciplines, technologies, and timescales are notorious. However, when it works the results are spectacular. To its credit ARL has taken on the challenge. ARL is one of the pioneers among major laboratories in the world in committing resources and building a program dedicated to network science on this scale.

Accomplishments and Advancements

Network Science Strategy

ARL's strategy has the following main elements: enterprise approach, network science driven by operational experience, expanding collaborations and partnerships, and focus on experimentation.

Although a good start, the strategy needs to be fleshed out and documented. There is evidence of promising starts in the enterprise approach and in expanding collaborations and partnerships, notably in the partnerships with universities in the Collaborative Technology Alliance (CTA) and the International Technology Alliance (ITA). Evidence on the same scale in support of network science driven by operational experience and a focus on experimentation was not apparent. Some of the main opportunities and challenges that the network science program faces can be encapsulated as the need for infusion of more operational experience and experimentation.

The Network Sciences and Information International Technology Alliance

A White House press release dated March 14, 2012, recognized ARL's work, noting the collaborative research done by U.S. and U.K. partners in the ITA to enhance information-sharing and distributed, secure, and flexible decision making in coalition operations.

One ITA project, Gaian, involves a distributed, dynamic, federated database. This project analyzes how to minimize time to reply to database queries, which may traverse an ad hoc network, or a hybrid of ad hoc and regular networks, where (sensor) data is distributed among the network nodes. Gaian appears to be a significant innovation. However, its results have not undergone comparative evaluation. There is a large body of work in both dynamic caching schemes and in mobile ad hoc networks, especially routing, areas of particular interest in the project. It is important to assess how well Gaian performs relative to the existing benchmarks, to determine whether the project's goals extend to understanding of the theoretical limits on performance, and, if so, the proposed scheme's relative performance. These matters were not clearly presented for review, and the subsequent discussion did not indicate that it has been sufficiently addressed by the project. A second ITA project focuses on the development of control algorithms to improve end-to-end performance by using "smart" data ferries.

Machine Translation

This project continues to reflect the high caliber of research that has been demonstrated over the past few years. The research successfully addresses the important Army requirement for effective translation of source language documents into target language and has produced a system that has been in use in the field. The ongoing research aims at improvements focusing on translation of heterogeneous text and dialects so that customized translation becomes feasible. The connections to universities (for example, Johns Hopkins University, University of Pennsylvania, Carnegie Mellon University, University of Massachusetts, and Massachusetts Institute of Technology), both directly and via multidisciplinary university research initiatives (MURIs), are excellent. The researchers are aware of external work, and so the research has been targeted to the Army niche that commercial developers are not addressing. The researchers are also addressing measurement of system effectiveness using training data. This project is an excellent example of the combination of research and systems engineering.

Image Processing

The context-from-imagery project is a good example of work that is customer-funded, includes significant components of 6.1 and 6.2 funding, has collaborations with universities, and intends to meet an Army need through a combination of science and engineering. The work characterizes face signatures in thermal and visible spectrum, with the goal to identify an image from a gallery of face images from thermal (long- and mid-wave infrared) imagery. Related prior works focused on near- and short-wave infrared, which are not as useful for nighttime identification of adversaries because active illumination is required. The technique is a composite of preprocessing, feature extraction, and recognition.

The results should specify type-1 and type-2 errors (false positives, false negatives). The basic research aspect of this project has been given short shrift, which may be due to the customer's request for quick results. Missing from the research component of this work is a sufficient descriptive framework, for example, a theoretical mapping from thermal visual space to physical space, calibrated with experiments, and with an underlay of causal explanations. This framework might start with investigating the mapping of a library of facial elements (for example, noses and eyes) and then developing extraction approaches based on a deeper understanding of phenomenology, which is likely to lead to different algorithms for different parts of the face. The chances of getting significant and major results would be greater if there was a greater investment in the research. The project's validation effort is honest. The publication in *Applied Optics* is good news.

Quantum Ghost Imaging

This work is progressing well and is supported by mission funding. It is also generating significant customer interest. The team has developed a good understanding of this imaging technique, including what is possible without harnessing the unique quantum correlation features, and what is additionally possible using quantum correlation such as some immunity to turbulence. The team already has shown experimentally that the large, incoherent aperture provided by the sun allows for higher resolution imaging over kilometer-scale distances than is possible when using the same optical hardware for conventional imaging. Because imaging, including imaging through turbulent media, is so central to the BED mission, and because this imaging modality is governed by significantly different constraints than is conventional imaging, it demands a thorough theoretical and experimental evaluation. It is rewarding to see that this is under way.

Atom-Photon Quantum Systems

This work is at a very basic research level and is exploring the ability to create remote entangled atomic states using optical fiber technology. This is very interesting work, involving leading-edge physics experiments.

Optical Communications and Networks

The optical communications and networks project, primarily supported by 6.1 funding, is well connected to universities and laboratories such as Lincoln Laboratory and the MIT Institute for Soldier Nanotechnology, and to industrial leaders such as BAE Systems and Raytheon. The effort is largely staffed at CISD, but there is collaboration with SEDD for ultraviolet (UV) sources. The goal is to realize unconventional optical communication systems, including UV non-line-of-sight (NLOS) and covert visible-light communications. An important application is intra-convoy communications in situations where jammers are being used to neutralize improvised explosive devices (IEDs), as well as when friendly RF communications are jammed.

Because the atmosphere is highly scattering in the UV range, the idea is to use the scattering to create NLOS communications. Realization of this idea has become feasible because of recent advances in devices, including UV light-emitting diodes (LEDs) and receivers. Prototype systems have been developed, for instance by BAE and Lincoln Laboratory, but these suffer from high costs, low availability, and low efficiency, all of which this project aims to ameliorate.

One of the team's accomplishments has been to model the propagation path loss as a function of scattering angle, distance, and pulse dispersion. For LEDs, transmission rates are in kbits/sec for distances less than 100 meters, and there is strong interdependence between data rates and distance. The team has performed short-range path loss measurements that have validated the model. The team is now developing UV laser technology for longer distances and higher data rates. The scattering model and Monte Carlo simulations of the model have been published in the *Journal of the Optical Society of America* (2011).

In related work the team authenticated a source-tagging protocol and presented its results during the Institute of Electrical and Electronics Engineers (IEEE) Photonics Society Summer Meeting in 2012. It is also extending point-to-point to bi-directional communications and relay nodes in the model. It has ongoing collaborative projects with Raytheon on tracking/tagging and quantum dots to shift wavelength, and with the Massachusetts Institute of Technology on optical filters. It is planning UV experiments for validation and transition to operational systems.

For visible light communications, the team has designed new constellations for modulation in transmitters. Because the design problem involves optimization over discrete variables, which has characteristics of a packing problem, the team adapted a novel Metropolis-like Monte Carlo algorithm originally developed at Bell Labs. A patent application on the constellation design has been filed.

This is a solid project that exploits nonconventional optical communications for the Army's future use. The overall vision is ambitious: to use synergy in capabilities across different segments of the spectrum. The project's progress has been good and is well supported by publications, peer reviews, and external visibility.

Because the concept of information transmission embedded in illumination systems is not new, the team should garner feedback from Army customers to evaluate system efficacy in one or more specific Army applications before advancing the basic research.

Opportunities and Challenges

Validation of the Network Science Vision and Strategy

ARL should approach the network science vision and strategy as a series of promising hypotheses, each of which remains to be validated. The following important question has not been clearly answered: Does the ARL research in network science reflect an understanding of the Army's requirements for the research? To answer the question rigorously, it will be helpful to understand how the results of the technical effort will translate to benefits for the Army and the soldier, and how the metrics will be used to quantify the benefits. Similarly, it will be helpful to have compelling examples of such benefits. However, because this research is in the early stages and there are no results to validate or compelling examples to review, the project should develop imaginary scenarios where the as-yet imaginary results of the research effort will have a beneficial impact. Of course, all of these factors need to be refined and iterated over time in parallel with the research's evolution.

The 2009-2010 assessment report contains a section titled "Challenges in Systems Engineering" in which it is noted that "it remains a significant challenge across CISD to ensure that even in relatively basic research programs a good understanding is formed about how potential systems that might be developed out of research might be deployed and used in real Army scenarios." The report adds that only a small amount of investment in the early stages of a program is likely to be repaid handsomely later in impact on the Army. These observations remain valid with regard to network science, given its large scale and ambitions, as well as its being in the early stages of the R&D continuum.

What Is NOT Network Science?

The labeling of certain projects as network science could only be justified by an extraordinary stretch of the definition of network science. Examples of such projects are machine translation and subjective logic using partial observation.

There are two points of view on this issue. With one, the "big umbrella" perspective, it does not matter how a work item is labeled as long as it is of high quality and value is generated. In fact, expediency argues for this approach, because network science is currently a hot topic, and it plays well with various clients to show that large resources are dedicated to it. With the other point of view, the lack of rigor in the definition and application of what exactly is meant by network science is likely to have unintended and detrimental consequences. For instance, genuine initiatives in network science may get squeezed out in the competition for resources. Also, the lack of rigor may have the pernicious corrupting effect of diluting standards of technical excellence.

ARL management should better define network science and its boundaries, articulate these definitions in a document, and apply them rigorously in categorizing work. Reconsideration of the entire computing research structure will also be associated with any redefinition of network science.

Cross-Genre Research

Although there was evidence of a desire to conduct cross-genre research, there were few examples of such research, suggesting that this aspect of the initiative is still in infancy. The best examples presented tended to have their focus in a single genre or, in exceptional cases, a combination of two genres. As the network science enterprise develops and matures, it will be increasingly important to identify compelling examples of research in composite networks that makes good on the potential that the network science enterprise strategy has staked out.

A good example of cross-genre research is the project on integrated analysis of social and information links, which aims to combine social-cognitive and information genres and was part of the Network Science CTA in partnership with various universities and ARL. This work combines data-mining techniques, estimation theory (maximum-likelihood, expectation-maximization, Cramer-Rao bounds), and optimization over graphs. The estimation and optimization techniques are standard in the repertory of signal processing and information theory in electrical engineering. The project has worked with 1.5 million tweets, which were collected in Egypt during the Arab Spring, to develop methods for reconstructing event timelines from real-time inputs. The inputs are, as might be expected, noisy, incomplete, and conflicting. The project is ambitious. Because noise, correlations, and dependencies abound, new insights might have come from the participation of research statisticians and mathematically trained social scientists, who have considerable experience in these problematic features. The results have been presented primarily in IEEE conferences and journals, with some ACM-sponsored forums as well. A broader discipline base would have been very desirable and would almost certainly have led to a broader array of approaches, methods, and tools to bear on the problem.

The above-mentioned project's connections to the Army's requirements were not articulated clearly; therefore, it is not apparent whether ARL is playing a sufficiently proactive role in defining and formulating the research problems in a manner that addresses the Army's requirements.

Work in the project on socially aware caching in disruption tolerant networks is solid. The results described are primarily from collaboration among university partners within the framework of the network science CTA. This work is positioned as cross-genre, combining social and communication networking. The intention is to use social interest and social relationships, such as roles and positions in the military, to increase the efficiency of caching in tactical mobile networks. However, intentions notwithstanding, the work is primarily mainstream networking, dominated by caching, together with various early-stage attempts to exploit the social dimension. This project would benefit greatly from the active participation and contributions of social scientists working closely with the electrical engineers and computer scientists who are already involved. It would also be valuable to have the Army requirements and perspective injected more compellingly into the problem formulation and the performance evaluations; only ARL can provide this perspective.

Diverse projects in the network science program, such as socially aware replication and caching and smarter middleware, are developing different techniques for getting the right information to the right people at the right time. These projects embody different measurement and modeling approaches to social networks, for example, graphs versus hypergraphs to reflect centrality or in-degree/out-degree. However, there was no presentation of what has been learned from the different approaches. For example, what are the merits and drawbacks of each approach? For what kinds of problems would given approaches be more viable? As these diverse projects progress, the value from syntheses becomes more compelling and the need for it more urgent. ARL should give such higher-level activity more attention.

Cross-Organizational Research

One of the laudable goals of the Network Science Enterprise is to combine the diverse disciplinary strengths from multiple directorates—in this case, CISD, HRED, and SEDD—to tackle a large problem with potential significant impact. This has yet to happen on any significant scale in the research and technical activities of the enterprise.

To counter any tendencies of staff to apply thinking that is bounded by the constraints of disciplines that are organized by current directorates, ARL should examine the current practices by which the network science programs and projects are organized. ARL should consider alternative means of organizing them, to determine the means that most clearly identify and empower technical leaders who will guide teams of staff drawn from across ARL directorates.

Trust

The presentations on the subject of trust, which is important in examining decision making, provided disappointingly incomplete results. The presentations quickly jumped to metrics and sophisticated mathematical models and their analyses. For a subject as amorphous and subjective as trust, it is important to work with data, and for data analysis and mathematical analysis to be mutually supportive and interactive. What data are being used to build the model, and do valid inferences from the data support the model? Examination of trust is certainly a case where real-life scenarios would be helpful at the outset to understand and define the problem before embarking on elaborate models and analyses.

Additionally, partnerships with experts in the social-cognitive field were not described. The assumptions, propositions, and the strategy for validation, all foundational topics, need to be carefully worked out. The research on trust, while important intrinsically and for its impact on multiple network genres, is an early-stage work-in-progress. Premature extrapolation can be harmful.

Social-Cognitive Sciences Research

The social-cognitive network researchers presented a compelling picture of the context of application in the tactical environment of untethered warfighter teams. How to provide effective support for distributed decision making in such environments constitutes a unique niche for ARL research.

In social-cognitive research more focus should be directed at understanding the variables that determine the right outcomes of effective decision making by untethered teams of soldiers. More attention needs to be directed at identifying the qualities of effective decisions. Identifying and specifying the key independent variables, as well as the dimensions of effective decisions, will reveal answers to some of the research questions being asked. An example of where this would be useful is the study that included 22 potential predictor variables for somewhat unspecified decision outcomes. This is where the modeling, measurement, and project methods should be directed initially.

An effective approach is to ground the research efforts in social/cognitive/behavioral theories of context-embedded cognition and social action. For example, much can be gleaned from decision-theoretical approaches, group/team decision making, and distributed decision making. Such research efforts have much to gain from exploiting the relevant research literature, which is extensive. Additionally, examination of the literature on fundamental social and behavioral sciences (theories, methods, and measurement methods) would be beneficial to the projects. The project managers should also seek out collaborators and visiting scientists who have strong backgrounds in experimental design, statistics, and creative approaches to measurements.

The planned work on increasing focus on group/social dynamics is appropriate, as are plans to expand attention to "influence" processes and improved outcomes of distributed decision making.

Shaping Behavior

The novel project on shaping behavior is part of the Social Cognitive Networks ARC component of the Network Science CTA. At its core, this project uses methods that combine crowd-sourcing with financial incentives—techniques also used by the presenter while leading the team that won the DARPA "red balloon" challenge. The goal of the project is to shape the behavior of a large number of people with small, carefully designed and placed incentives.

This is a topical subject with great promise for the Army and society in general. The project is in the very early stages of development. This work represents a rare instantiation of the socio-technical approach espoused by the program. It does so by elevating attention from individuals to social-level variables (the hypotheses are all based on groups and social behavior, as opposed to individuals and person-specific roles). The work relies on large networked unstructured communications, as opposed to pair-wise connections. It is not yet clear how this work may be applied in Army-relevant distributed cognition, decision making, and performance, which is an appropriate future direction warranting exploration.

A centerpiece of this presentation was a sketch of the mathematical modeling and optimization for the design of incentives and predicting responses. The mathematical development was fairly standard and at a very early stage.

"Big Data" Research in Network Science

ARL's and CISD's management clearly expressed their cognizance of the problems associated with flooding of data ("drowning in data") and the opportunity cost of not being able to exploit available data and information to benefit the soldier. Of the top five "big ideas" in the Director's presentation, two are specifically tied to this topic: "decision science" and the tactical cloudlets element of computational science.

However, ARL did not convincingly demonstrate that it has garnered and organized into a comprehensive understanding all aspects of this challenging problem. The research in academia and industry in this area has now gained considerable momentum. This effort has brought together theory, algorithms, and systems, and it has roots in several disciplines, including computer science, mathematics, statistics, and operations research. Machine learning, together with data mining, knowledge discovery, and social network analysis, which were until quite recently obscure branches of artificial intelligence and statistics, are now flourishing within the framework of "big data." The solutions are no longer static, and the data populating the databases are derived by and drive sophisticated analytic algorithms. The information is being derived from humans and machines (sensors), as well as from huge stores of information on the Web. The knowledge that is mined will have to be comprehensible to humans, and humans making decisions will have to be able to actively question the assumptions behind the knowledge extraction process.

ARL should be an important driver in this movement, or at least a judiciously selected subset of the movement. ARL has the advantage of having abundant data, the lack of which is often the bane of researchers. ARL should use this advantage to attract high-quality external researchers to collaborate on problems of interest to it. Additionally, it should partner with other Department of Defense (DoD) agencies that are actively involved. For example, the Air Force Office of Scientific Research (AFOSR) sponsors a high-quality program in network science that is viewed as a leader in fundamental research in large data analysis.

Network Science CTA and Networks and Information Sciences ITA

Four university professors who are participating in the Network Sciences CTA and ITA provided presentations describing high-quality research that has been, and continues to be, conducted by teams consisting of academic and ARL researchers. The research presented was driven and conducted by university collaborators in the consortia with contributions from ARL researchers that did not appear to be on par. Although the level of connection was uneven across the consortium projects, the level of connection to the Army's needs, the understanding of the Army's needs, and the reflection of constraints from the field in the problem formulations were lower than should be the case.

It is important that ARL find solutions to the challenging problems of less than ideal ARL involvement in the consortia's research and less than ideal connectivity of consortia research with Army needs. Any solution will require ARL to be more proactive, even aggressive, with its academic partners in translating and then inserting the perspectives from the field into the problem formulation. The collaborators from ARL will need to match their scientific and technical abilities, as well as their status in their research communities, with that of their university partners.

To be clear, the question is not whether ARL is deriving value from the consortia, for which the answer is definitively in the affirmative, but whether the derived value is providing the impact that ARL desires. There is undoubted value from being aware and being involved, at any level, in high-quality research. However, if this value is to be more than educational, then ARL should perform an in-depth investigation of the means and desired outcomes for extracting value from these collaborations. Such an investigation should consider two models for extracting value from collaborations, discussed below under the section "How to Extract Greater Value from University Collaborations."

Overall Technical Quality of the Work

The technical quality of the work was assessed according to several criteria posed by ARL. The first criterion asks whether the scientific quality of the research is of comparable technical quality to that executed in leading federal, university, and/or industrial laboratories both nationally and internationally. The answer is generally affirmative for the network science program and for the two divisions in CISD that are intimately involved in the program, the Information Sciences Division and the Network Sciences Division. Exceptional expertise exists in selected areas, such as machine translation, image processing, anomaly detection, and optical communications and networks. There are extensive collaborations with external research entities, especially universities in the United States and United Kingdom, and leading industrial laboratories. Of concern is the extent and nature of the role of ARL staff members in the conception, formulation, and execution of the research in the collaborations; a related concern is whether the research's impact on the Army will be sufficient.

CISD has demonstrated greater awareness, at both the managerial and staff member levels, of the desirability of publishing in peer-reviewed and archival journals. An increase in publications in high-quality journals provides evidence that this awareness is slowly having an effect. However, evidence of staff participation in scientific and professional societies is lacking. Multifarious benefits would accrue from such involvement, such as acquiring peer status in external collaborations and collecting information on the state-of-the-art in technical developments that would be useful in expanding external collaborations. Recognition from professional societies would help recruiting and retention. A basic and quite easy step to take in this direction is organizing and hosting workshops. (This has already been done in varying degrees across organizations.) More effective use of enticements that ARL has at its disposal, for example, its extensive data sets, should be used to encourage participation by universities and industry in ARL-led meetings and workshops located in the ARL facilities and also in high-quality conferences.

The second criterion applied during the assessment is whether the research program reflects a broad understanding of the underlying science and research conducted elsewhere. The answer here varies with individual projects, and generally the projects considered to be exceptional are the ones with the best understanding. It should be pointed out that network science is a fledgling area of scientific and engineering research. It is to ARL's great credit that it has embraced the subject, and ARL is in many ways a pioneer among major laboratories in doing so on the scale that it has. However, not a great deal of basic science and research is being conducted elsewhere to compare to.

Nonetheless, there is room for improvement. With the exception of research on shaping behavior, most of the work on distributed decision making involves study of two-person teams, which does not live up to the aspirations of the network science program. After all, research on two-party interactions has been going on for decades, well before contemporary digitally networked media became available. Research needs to reflect the size of distributed communication and decision-making teams in the compellingly depicted tactical environment of untethered warfighters. Providing effective support for collaborative activities in such environments represents a unique niche for ARL research.

There is substantial research in areas adjacent to network science, if not directly within it (depending on its definition), that the ARL program should better understand and embrace. In areas such as "big data," including data mining and anomaly detection, machine learning has taken on a large, even dominant, role. ARL has no tradition of research in this area; a special effort will have to be made to remedy the situation.

The answer to the third assessment criterion, that is, whether facilities and laboratory equipment are state of the art, is largely a solid "Yes." An important exception is in the cloud computing area.

The fourth assessment criterion asks whether the research team's qualifications match the technical challenges. The match is good, but it could be substantially enhanced by remedial action.

The fifth assessment criterion asks whether the research reflects an understanding of the Army's needs. The answer is that the research generally reflects the Army's needs. However, in several cases, the research's potential to successfully address specific Army needs was not clearly defined.

The sixth assessment criterion deals with the crafting of programs that employ the appropriate mix of theory, computation, and experimentation, which ARL management has clearly identified as an objective, for example in its network science strategy statement. However, the results are uneven. This criterion is coupled to topics such as feedback, validation, and the desirability of greater systems engineering. Not surprisingly, the less established the research topic, such as the integration of social-cognitive genre with other genres, the greater is the gap between the ideal and the current states of integration of theory, computation, and experimentation. This is not surprising, because time is needed to build up a research area. Experimentation is of such overall importance in the conduct of ARL research that it should be an independent subject of research, for which there is precedence. For example, experimental design and exploratory data analysis are established research areas. Because statistics is the natural home for these topics, the involvement of professional statisticians would be helpful.

The seventh assessment criterion asks whether especially promising projects could be transitioned ultimately to the field. Several projects have been identified as promising for ultimate transition. There is more that ARL could do, however, with respect to the selection of research focus in the CTA and ITA. Similarly, systems engineering considerations also can be applied from the inception of projects. It would be important to register, track, and quantify the impact of transitions from research to the field.

ARL remains extraordinarily responsive to the ARLTAB's recommendations. This responsiveness has not diminished over time. Indeed, it has sharpened. It is gratifying for the ARLTAB to see effective responses to recommendations within a year's span; this is a characteristic of all the CISD units assessed.

COMMON THEMES

The following two sections discuss common themes that cut across the Network Science Enterprise and CISD.

Opportunities and Challenges

Cloud Computing Research

The investment in cloud computing research in CISD, and generally in ARL, appears small. ARL should consider the merits of expanding its work in this area. A particular concern is that tactical applications of cloud computing, such as tactical cloudlets, are particularly demanding variants, which should be tackled only after achieving mastery of the science and technology of basic cloud computing. ARL should consider making available Hadoop clusters to jumpstart research from the bottom up. These clusters, supported by the necessary software engineering knowledge, should be used as platforms to develop machine learning, data mining, search, social networks analysis, and various other applications. Other dimensions to cloud computing, especially for tactical applications, call out for collaborative research, for instance cyber-security and special communication network protocols and resource scheduling to handle near real-time applications (of obvious importance in tactical applications) within the geographically dispersed context of cloud computing. Extensive use of caching will almost certainly be necessary, and mastery of the enabling algorithms will be prerequisite. From a networking perspective, cloud computing integrates processing, storage, and communications, and the research challenges are many. When realized, cloud computing has the potential to give the soldier a significant technological advantage.

Cyber-Security Research

Cyber-security science has been positioned as starting from a relatively small base with intentions to be a growth area. For instance, CISD lists cyberdefense as one of its "top 5 future big ideas." Management of the Network Sciences Division (NSD) has deferred questions of strategy for cyber-security research to the next year, because that strategy has not yet been defined. These decisions are appropriate.

A significant amount of thought and effort will be needed to formulate a strategy and an action plan. The topics in cyber-security are expansive, and the competition for high-quality researchers is fierce and growing. ARL should adopt a targeted approach to reach Army-specific goals in order to maximize impact on Army cyber-systems. This can be achieved through careful planning that selects problems that are difficult but can feasibly be addressed by a relatively small group of researchers.

Considerable capabilities exist in other DoD agencies in such areas as cyptography and malware analysis. Without extensive knowledge of the work by others in these areas, CISD risks duplication of effort. CISD is working in the areas of intrusion detection systems and alert management. There is a considerable body of prior work and a sophisticated and broad commercial market for technologies in these areas. It does not appear as though there are important research topics to be mined in the areas of intrusion detection systems and alert management.

There are many research questions related to advances in data leak prevention that require new insights and technical developments that are not shackled by the biases of unimaginative approaches to authentication and authorization. ARL may wish to consider and develop this area of research. Scalable deception is an underdeveloped technology that is directly responsive to data leak prevention and that is fraught with challenging research problems. Successful work in this area would respond to the congressional mandate to prevent data leaks from sensitive DoD sources.

Also worthy of further consideration are the topics of "science of security" and "security metrics." These, together with the topic of insider threat, involve research problems that a small number of high-quality researchers in ARL can address with potentially high impact.

Overall Quality of the Work

The Advisability of Broadening the Discipline Base and Strengthening Fundamental Research

There is great value in developing a broader base in the disciplines at ARL. Certain disciplines, such as computer science, are obviously greatly needed and also fairly well represented. Yet other disciplines are under-represented or not represented. One example is statistics. Several areas are closely connected to statistics, including machine learning and statistical computing. Statisticians characteristically support research across disciplines and organizations. Social scientists, especially when trained in mathematics and quantitative analysis, have the potential to significantly lift the quality of several ARL projects, especially in the social-cognitive area. There is ample scope of projects for mathematicians at ARL. Consider, for example, the value of mathematical formulations of the information network and its interactions with adversaries, mathematical formulations of features in data, and secure multi-party computations. It is not a big jump from mathematical formulations of structures to algorithms that allow fast computations from data. Given the competition for algorithm specialists, a possible path may be to hire mathematicians and train them in algorithms.

ARL and CISD would benefit from a bigger commitment to fundamental research. There is considerable room for improvement in fundamental understanding of problems and in advancing fundamental scientific knowledge. Especially in theory, this dimension of research seems to have been conceded to the university collaborators. If true, then this would be a mistake. It is essential for ARL to be equal partners in all dimensions of any research collaboration to extract full value.

ARL Researchers' Interactions with the External Technical Community

ARL management and staff members have already taken to heart prior ARLTAB recommendations for proactive engagement in the technical community, especially outside the Army and DoD. Continuing that progress is recommended.

External Recognition

ARL should approach external recognition in professional societies in a coordinated, systematic way that is driven by appreciation of the multi-dimensional value that would be generated. Consider the IEEE, which is representative of professional societies. It has three membership levels: member, senior member, and fellow. IEEE requires membership for a minimum number of years at one level to be eligible for membership at the next level. ARL should consider this hierarchy as a pipeline, and it should provide incentives for its staff to advance to the next stage and to provide mentoring and other forms of assistance. This process could be replicated across key societies.

Professional Activities

Professional activities may be conducted on a stand-alone basis (for example, ARL would organize and host meetings of broad interest, inviting high-quality, high-visibility external speakers); in partnership (for example, with other federal agencies); or with the support of the professional societies and the organizers of high-quality conferences. Examples of activities include organizing workshops, seminars, and conference sessions on topics of special interest to the Army. Being an organizer will make it easier for ARL to accommodate other ARL researchers in the program. Locating such meetings in ARL facilities has additional value.

Centers of Excellence

The National Science Foundation (NSF), among other agencies, funds many centers of excellence. ARL should consider engaging such centers to help drive its research program. The NSF and the AFOSR, for example, typically conduct a stream of workshops, research programs, and other engagements with specific technical communities. ARL could take advantage of these meetings and workshops to refine the formulation of research problems and to better position itself to press its agenda in work with consortium partners.

Publications

CISD should build on its recent progress with publications by pressing ahead with its program to increase the quality and quantity of publications in peer-reviewed, archival journals. ARL and CISD should take a systematic approach, starting with the composition of an inventory of all journals and conferences that are deemed to be top tier in various respective fields, a tracking system of publications by divisions, and a system of incentives and rewards for staff members.

Online Courses

Online courses are becoming widespread, and quality online learning opportunities are increasing. ARL and CISD should consider exploiting this nascent practice and technology. Staff members would benefit from completing courses in computer science, artificial intelligence, and programming taught by experts from top schools. Conversely, ARL should consider offering select courses for ARL staff and possibly for staff at the extended DoD community.

How to Extract Greater Value from University Collaborations

ARL may not be realizing top value from its investments in the university consortia. Quite often the research itself is of high quality, but equally often it is more suited to goals posed by academia and offers less than what might be possible for ARL and its staff, the Army, and for advancing the likelihood of transitioning to the field. ARL is encouraged to examine the means for extracting greater value from such collaborative consortia. ARL should consider two models for extracting top value from the university collaborations, and these are not mutually exclusive. In the first model, ARL researchers are peers to their academic partners in the research, so that the typical, frequent, face-to-face interactions that might occur among university collaborators in the same corridor also occur across the university-ARL boundary. For this model to apply the ARL researchers need to possess the same depth and stature in the technical area as the university researcher, and, from what has been observed, this is typically substantial. In the second model, ARL researchers perform the invaluable function of translating Army requirements and educating the academic partners on field constraints and Army needs. In this capacity, ARL helps to transform academic problems into no less challenging problems that are also of interest to the Army; that is, ARL staff act as links in the bidirectional feedback loop between academia and the laboratory.

Both models have value for ARL, and both have demanding requirements. For the first, collaborating researchers from ARL need to have as deep a fundamental understanding of the science as the university collaborator. Other characteristics also need to be aligned. For example, the publication records need to be comparable, as do external recognitions. In the second model, for the field constraints to be persuasive and for clever, responsive problem formulations, the intellectual depth also needs to be comparable on both sides. In several respects the second model is more demanding. ARL should decide on a model and an implementation strategy.

Staffing

Staffing gaps, especially in connection to disciplines, were mentioned above, as have staffing issues in building the embryonic cyber-security research area and in balancing research and HPC infrastructure support. Another noteworthy case is the under-staffing of the Information Sciences Division (ISD) in CISD. Given the pulls on ISD from diverse direction, including machine translation, robotics, and customer-driven activities, the residual resources are insufficient for addressing the problems of data, which should be at the core of ISD's research agenda. Beyond undercutting a key technical area, the understaffing of ISD has other consequences. The comparative data across divisions on publications in archival journal, after normalization by the number of Ph.D.'s in the divisions, was notably lower for ISD, which may be due to under-staffing. CISD should consider the staffing issue in ISD.

Role of Patents to Give Impetus to the Culture of Innovations

Patents are an important part of the culture of industrial laboratories and increasingly of universities. The role of patents in ARL's culture appears to be relatively small, which may be how it should remain. However, when properly managed, patenting in the mind-set of industrial researchers enhances the culture of innovation. Major patents and their inventors are recognized universally, with significant benefits to the home organization. Financial rewards are a major incentive. Having the requisite staff to manage the legalities and processing implies increased overhead. It is good practice for organizations such as ARL to periodically review their policies pertaining to patents, the potential for collaborating and pooling resources across sister government organizations, and the methods of communicating policies to staff.

Director's Strategic Initiative and Director's Research Initiative

ARL is to be complimented for establishing and maintaining its Director's Strategic Initiative (DSI) and Director's Research Initiative (DRI), which are nurturing a bottom-up approach to the selection of research problems. The breadth and relevance of the 16 research topics currently supported by DRI are indeed impressive. If it is not already being done, tracking and analyzing the history of these projects may be useful in guiding research management.

The CISD management and research teams continue to be responsive to ARLTAB's recommendations. CISD has made significant changes in strategies and research processes that alleviate problems on which the ARLTAB commented in the past. Also noteworthy is the care and attention that CISD's management, starting with the Director and the division chiefs and extending to staff members, have given to planning the panel meetings and preparing the material for presentations in the panel reviews and posters. The ARLTAB appreciates the year-to-year improvement in the proceedings. All of these developments are evidence to support the belief that ARL is receptive to feedback from the ARLTAB. Such receptivity is a hallmark of a dynamic institution that constantly strives for improvement in the pursuit of technical excellence.

3

Human Research and Engineering Directorate

INTRODUCTION

The Soldier Systems Panel visited the Human Research and Engineering Directorate (HRED) at Aberdeen Proving Ground, Maryland, on May 15-18, 2011, and June 19-22, 2012, and examined written materials provided by HRED. During those visits, the panel received briefings on aspects of HRED work, mostly in the 6.1 (basic research) and 6.2 (applied research) categories. This chapter provides an evaluation of that work, recognizing that it represents only a portion of HRED's portfolio.

Overall Organization

Research in HRED is currently organized around two large, major laboratory programs: the Human Dimension program and the Simulation and Training Technology program. This chapter is organized along the seven research thrusts within the Human Dimension program:

1. *Sensory performance.* The sensory performance thrust is concerned with the sensory performance of individual soldiers and small teams as they carry out tasks that are critical to their military roles. The research is focused on the interaction of the dismounted soldier with his/her equipment and environment. Examples might include studying the impact of protective gear on the perceptual ability to detect, identify, and localize threats while also maintaining general situational awareness and the effects of devices intended to enhance natural sensory capabilities (for example, night vision devices).
2. *Physical and cognitive performance interaction.* Soldiers carry heavy (more than 100 lb) loads for extended periods of time (for example, 72-hour operations), often over rough terrain and while being inundated with large amounts of information. The physical and cognitive perfor-

mance interaction thrust investigates the effects of the interaction of cognitive load and physical and cognitive stress on the soldiers and on their performance.

3. *Translational neuroscience.* As Army operations enter battlefields that are more dynamic and complex, it will become increasingly critical to build translational capability into the design and development of systems that capitalize on the soldier's neurocognitive abilities to meet the demands of these environments, ensuring mission effectiveness and maximizing soldier survivability. The goal of the translational neuroscience thrust is to enable system designs that are consistent with brain function, taking into account its limitations and exploiting its potentials, to maximize soldier performance.
4. *Social-cognitive network science.* The soldier on the modern battlefield is increasingly part of a network of humans and machines in rich communication with one another. Decision making is increasingly distributed and dispersed. This reality poses a set of questions about presentation and consumption of information that is the domain of the social-cognitive network science thrust area. These human and human-machine questions are related to questions about the structure and function of networks more generally. Consequently, the primary discussions of this thrust area will be found in the chapters of this report that discuss the Computational and Information Sciences Directorate and crosscutting research areas.
5. *Human-robot interaction.* An increasing number of unmanned, robotic systems with various degrees of autonomy are being fielded. Humans need to control these devices and to plan their activities. In addition, humans receive and interpret input from robotic devices. Optimal (or even merely adequate) use of these devices requires an understanding of human-robot interactions. Such interactions were reviewed as part of the autonomous systems enterprise work within the Vehicle Technology Directorate. Accordingly, this research thrust is examined in the chapter covering the autonomous systems enterprise.
6. *Human-systems integration (HSI).* The human-systems integration thrust assesses the ability of soldiers to work with new systems prior to their deployment. Much of this work is HRED's contribution to the Manpower and Personnel Integration (MANPRINT) program, a formal Army program with the goal of ensuring that most if not all human dimensions are accounted for in the design, development, procurement, and life cycle management of all Army materiel systems. A leading HSI tool for MANPRINT work is IMPRINT (the Improved Performance Research Integration Tool), a modeling system based on discrete event simulation. IMPRINT is one of several tools used by HRED to produce early, cost-effective evaluation of constraints that can inform the determination of requirements for new systems.
7. *Opportunity-driven human factors research.* HRED scientists have the expertise to address novel problems and questions as they arise in the field. The opportunity driven research thrust is the home for research directed at specific problems faced by soldiers in the field or by equipment development programs. Opportunity driven research is typically customer funded and occurs throughout HRED.
8. *Simulation and training technology.* Simulation and training technology is a major laboratory program with five research areas: adaptive and intelligent training technologies, synthetic environments, immersive learning, training applications, and advanced distributed simulation. Training applications are domain-specific research areas and include medical, dismounted soldier, and ground platform training. The laboratory's primary physical location is in Orlando, Florida, and it has a strong relationship with the Institute for Creative Technology at the University of Southern California.

DIRECTORATE-WIDE THEMES

The HRED mission in 2012 can be separated into three large parts. First, there is research intended to improve soldier performance in the future. This work involves research on humans and their capabilities and on human-machine interactions. The most basic and speculative aspects of HRED research fall within this area. Second, because HRED is a center of human factors expertise, it is in a position to evaluate the integration of soldiers and systems within the Army. Importantly, it can help to shape requirements during the acquisition process to ensure adequate consideration of the human dimension. Research within this customer-service aspect of HRED's work is devoted to improving assessment tools such as IMPRINT. Third, with the inclusion of the Simulation and Training Technology Center (STTC), HRED is assuming an important role in improving the training of soldiers through the use of artificial training environments. STTC's research involves the development of new simulation tools such as generalizable frameworks that speed development of specific training and simulation tools.

The wide scope of these research areas poses a central challenge to HRED. Given limited resources of time, personnel, and funding, what specific topics should HRED tackle? Of course there are always more topics of interest than can be addressed. Like other research labs, some of HRED's choices are driven by funding. The ability to perform basic research is dependent on the continued availability of basic research (6.1) funds from the Army.

The Soldier System Panel was charged to assess the scientific and technical quality of the Army Research Laboratory's (ARL's) research and development related to its soldier systems mission and to examine how HRED's work compares to similar work being done externally. As a result, the panel focused its attention on questions about the HRED research that might have an impact on the broader scientific community. This focus is not meant to denigrate specific work performed by HRED to solve specific problems with specific systems. Such work is a vital part of HRED's mission, and the challenge lies in producing synergies between basic research and specific problem-solving work. By its nature, not all basic research will produce results that can be transitioned to Army-specific problems in the field. However, basic research can produce new classes of solutions. The answer to a customer's human factors question may not be publishable in the scientific literature. However, a workforce, encouraged to think as basic scientists, may find that a specific task raises new topics for basic research. HRED's leadership is tasked with steering a course than enables HRED to strike a balance between its basic research and specific, problem-solving missions.

CHANGES SINCE THE PREVIOUS REVIEW

The appointment of a new HRED Director in January 2011 has provided stability to HRED management. A senior scientist in neuroscience has also been appointed, which represents an important milestone in the development of the translational neuroscience thrust in HRED.

In terms of facilities, the most significant change has been the opening of the Soldier Performance and Equipment Advanced Research (SPEAR) facility. This facility, with its instrumented obstacle, cross-country courses, and biomechanics laboratory, gives HRED important new capabilities for measuring soldier performance under realistic conditions. With regard to staffing, the recruitment of early-career, clever scientists at a time of little or no growth is commendable.

A previous report noted that some research areas were governed by a clearer scientific vision than were others.[1] During the past 2 years, HRED leadership has made it a priority to develop a clear vision for each of the research thrust areas. Although a work in progress, this decision reflects a noticeable change since the previous review.

There is increased evidence of good contacts with the broader scientific community through presentations at meetings and publications in the peer-reviewed literature. Moreover, the quality of presentations has generally improved, suggesting a steady improvement in the research. Additionally, the background and hypotheses motivating individual research projects are more clearly stated, and the connection to scientific research beyond HRED is documented more frequently.

The presentation of data and the use of statistics have been improved, although statistically underpowered studies and questionable statistical analyses remain. A more subtle issue is the distinction between statistical significance and scientific or practical significance. It is possible to have statistically reliable results that are, nevertheless, of little consequence, which has not been clearly recognized. At the directorate-wide level, this represents a locus of intervention, where senior, experienced researchers may be able to provide useful mentoring to less seasoned staff.

Quite a few projects seem to be stand-alone studies; that is, they are initial experiments of what would be a series of studies if substantial publication were to be the result. This may also be a place where scientific mentoring by senior researchers would be helpful. The data being gathered by the opportunity-driven research represents an opportunity to close the loop between basic and applied research and soldier system problem solving. To support this effort, the hiring of more senior, experienced scientists should be complemented by the better use of available field operations reports, and even medical reports, to prioritize studies and plan how they should be executed.

SENSORY PERFORMANCE

Changes Since the Previous Review

The most notable positive change in the sensory performance area has been the progress toward a strategic vision. The most notable challenge is the need to develop a program of basic and applied research that implements such a vision.

Accomplishments and Advancements

In its past two reports, the ARLTAB stated that the sensory performance area lacked clear direction.[2,3] Although the direction is still not entirely clear, progress is visible. The sensory performance thrust focuses on the individual soldier and on small teams. The thrust's main areas of work are in auditory and visual performance. There is a clear understanding that HRED's work should concentrate on soldier-critical sensory performance capabilities, including such issues as the ability to detect threats (e.g., improvised explosive devices [IEDs]), interpret the intentions of bystanders, and communicate without

[1]National Research Council. 2010. *2009-2010 Assessment of the Army Research Laboratory.* Washington, D.C.: The National Academies Press, p. 33.

[2]National Research Council. 2010. *2009-2010 Assessment of the Army Research Laboratory.* Washington, D.C.: The National Academies Press, p. 43.

[3]National Research Council. 2010. *2007-2008 Assessment of the Army Research Laboratory.* Washington, D.C.: The National Academies Press, p. 34.

being seen or heard. All of these goals should be accomplished in a sensory environment shaped by the soldier's gear (e.g., protective equipment).

The sensory performance portfolio consists of three types of research: (1) basic and applied research to understand sensory-perceptual capabilities, (2) human factors research to understand the impact of specific equipment on performance, and (3) developmental work to examine new approaches to sensory enhancement. Although the primary focus is on the optimization and enhancement of our perceptual capabilities, efforts are under way to develop methods of manipulating perceptual information, reducing or altering the perceptual information available to others, and thus allowing the soldier to operate with stealth capability. More recently, a refinement of this vision has appeared under the rubric of "owning the environment." Here, the emphasis is on use of stealth and deception as force multipliers. Through better equipment and better understanding of the perceptual capabilities of friend and foe alike, it may be possible to reduce threats to our soldiers while increasing their effectiveness. Although in an early stage of study and not yet giving rise to a specific research program, the concept of using stealth and deception to own the environment is an interesting direction for work in sensory performance. The sensory performance group's state-of-the-art Environment for Auditory Research (EAR) facility enables very sophisticated basic and applied auditory research.

The clearest examples of accomplishments in this area are human factors projects, such as work on the effects of military headgear on auditory performance. The wide range of helmet designs results in differing effects on the ability to detect and localize sounds. The EAR facility is well-suited to conducting research in this area, with the eventual goal of developing a predictive model that can influence future designs. This is a good example of a scientific project that could transition back to an evaluation tool like IMPRINT.

Another application of basic sensory science to an Army setting is the work on the ability to recognize different types of small arms by the sounds they make when fired. Although it is less clear whether this application will result in a generalizable piece of new scientific knowledge or an obvious transition to a practical application, soldiers will benefit greatly from improved ability to identify what is being fired in the vicinity.

Further examples of potentially useful human factors and assessment work within this area involve the comparison of different interfaces for a missile defense system and the study of oculomotor indicators of workload in the complex setting of a helicopter cockpit.

Opportunities and Challenges

Impressive progress has been made in the sensory performance area, but many significant challenges remain, including the following:

- The vision for sensory performance research is still a work in progress. For example, it is not clear whether "owning the battlefield" will replace earlier formulations or simply be added to the mix. HRED understands that its strategic vision in this regard is still evolving.
- Once in place, there needs to be a major effort to align the strategic vision with the work. There will be legacy projects that will end, and new research areas that will need to be developed.
- HRED's best work falls under the category of human factors evaluation. Although the evaluations of various off-the-shelf systems can provide an important service to the Army, as a general rule, they do not connect to progress in basic science. Given HRED's laboratory capabilities, one would like to see a research loop in which an evaluation task raises basic science questions. Those basic science questions should be studied in a rigorous manner, and the results should be

published in the open scientific literature whenever possible. In addition, those results, along with other findings from the basic science community, could be used to support more applied research to shape the development of the next generation of systems. All of the pieces of this circuit can be found in the sensory performance area, but, as noted, most of the success seems to lie in the first—evaluation—step, and there is little evidence of programmatic coordination between these steps.

- Some of the work, although well-intentioned, is not methodologically strong. For example, an interesting project on IED detection made good use of anecdotal accounts from soldiers in the field but never managed to convert those accounts into hypothesis-driven research with the statistical power to draw conclusions.
- Several projects have not been influenced by the relevant scientific literature. For example, a project on detection of muzzle flashes in the periphery would have benefited from more consideration of the vast clinical literature on detection of targets in peripheral vision.
- Overall, the empirical work is dated; that is, it is more typical of psychophysical work conducted 20 years ago. There is room in the basic science work for greater use of naturalistic stimuli and field data derived from opportunistic research, more consideration of the role of attention, and more modern theoretical constructs.
- The state-of-the-art EAR facility continues to be underutilized and lacks a world-class program of basic research. Here, and perhaps elsewhere (for example, in the area of visual perception and attention), it might be worth considering bringing in a senior researcher from academia on a temporary basis (e.g., via the Intergovernmental Personnel Act [IPA] mechanism) or permanent basis to help shape such a program.

The previous four items present an opportunity to reach out to the broader community in studies of sensation, perception, and attention. Funding vehicles such as the Army Research Office (ARO), multidisciplinary university research initiative (MURI) system, or the Collaborative Technology Alliance (CTA) vehicle could be used to foster productive relationships between those working in the basic sensory performance world and those interested in application of research to Army problems.

Overall Quality of Research

In general, the strongest work in the sensory performance area is customer-driven evaluation of equipment. The group is not yet a force in basic or applied research on sensory performance.

PHYSICAL AND COGNITIVE PERFORMANCE INTERACTION

Changes Since the Previous Review

There have been two notable changes since the previous review. The first is an evolving focus on the interaction of physical and cognitive factors in soldier performance. The second is the completion of the SPEAR facility, giving this research thrust a particularly strong set of tools for research. Current physical and cognitive performance interaction research consists of three thrust areas aimed at quantifying the effects of soldier equipment on physical and cognitive performance and their interaction: (1) Developing new measurement methods/devices for collecting cognitive and physical performance data in operationally relevant environments, (2) Employing traditional and new metrics to quantify the effects of soldier equipment on soldier performance, and (3) Applying the new metrics to understanding the effects of physical and cognitive load on soldier performance over longer duration activities.

Accomplishments and Advancements

Biomechanics has long been at the core of HRED's mission to examine the physical limits on soldier performance. HRED has developed a very strong set of facilities for studying soldier performance under various approximations of the real-world situation. Field studies are extremely valuable for their realism but are not well suited to well-controlled, basic research studies. For such studies, HRED has multiple research venues for examining the interaction of physical and cognitive stress on soldier performance. These facilities include first-person gaming facilities for isolating the effects of cognitive stress (such as the C4ISR laboratory) and a state-of-the-art biomechanics laboratory for investigating the physical effects of load. The Tactical Environment Simulation Facility (TESF) is a fully immersive simulator that allows research subjects to walk through and interact with their environment. The M-Range Shooter performance research facility is a live-fire range capable of providing real-time high-fidelity data on marksmanship performance. The newest facility, the Soldier Performance and Equipment Advanced Research (SPEAR) facility (opened in the summer of 2012) combines HRED's biomechanics laboratory, outdoor obstacle course, and a WiFi-networked, 2.5-mile cross-country course that goes through the woods of Aberdeen Proving Ground. This array of facilities allows HRED a high level of experimental control if needed (i.e., in the C4ISR or biomechanics lab) and a high level of operational relevance if needed (i.e., in the SPEAR).

With these tools in hand, the task becomes deciding where to focus research effort. HRED has identified a niche that it is, perhaps, uniquely suited to study—the interaction of physical and cognitive stress on soldier performance. HRED notes that not only are soldiers required in modern missions to carry very large physical loads (in excess of 100 pounds), but also they need to do so in ambiguous, yet threatening environments (such as patrolling potentially hostile city streets), while making split second, life or death decisions. Few research facilities are capable of supporting the study of this combination of factors; this capability is a strength of the physical and cognitive performance interaction research thrust.

Because considerable effort has been devoted to the development of these new facilities, it is not surprising that a significant portion of recent work has involved validation of devices including the omni-directional treadmill, the force plate treadmill, and aspects of marksmanship measurement.

During the past year, the group has started to demonstrate how it can use its tools in work that moves beyond validation. For example, if the laboratory has multiple ways to test the effects of physical and cognitive load, then does it matter if you test in the real world or in a simulator? The answer from one recent study is that this does matter and needs to be taken into account. World-class, one-of-a-kind facilities notwithstanding, another recent project showed how to make clever use of off-the-shelf tools such as the Microsoft Kinect sensor and the Sony Move system.

The work on hand carriage of load is a good example of the sort of classic "bread and butter" biomechanics work at the center of HRED's mission to understand the demands placed on soldiers.

Opportunities and Challenges

The physical and cognitive performance interaction research group has a vision for a research program that exploits its unique and impressive set of available facilities. The challenge chiefly lies in implementing that vision while responding to more traditional requests for support, which is recognized in the current research plan. The group plans to continue to employ traditional methods to quantify the effects of specific equipment on soldier performance. At the same time, it plans to develop new methods to study both cognitive and physical data and to do so in the more realistic environments that are now available. The goal is to use these new metrics to study and understand how physical and cognitive loads impact physical and cognitive performance in activities over an extended period of time.

This group is in a good position to make progress in the next few years. It faces some challenges, many of which it has identified itself:

- Because the world of possible research is vast, the group needs to be strategic in choosing its work. Possibly these choices should be driven by closer contact with infantry and armor operations. More generally, continuing broad collaborative contacts with both the Army and the broader scientific community will help to assure that the research program does not become insular.
- It is obvious that research of this sort requires a steady stream of human volunteer participants. Less obvious is where those human subjects will come from. This is a general HRED issue, but it is particularly salient in the physical and cognitive performance area, because the specialized equipment and facilities require that subjects go to Aberdeen. The group has made some excellent efforts at outreach to different groups of potential subjects, in effect trading participation for benefits that those groups can obtain from the use of HRED facilities. Nevertheless, this is a chronic problem that deserves continuing thought by HRED management.
- Related to both of the preceding points, the group should continue its efforts to study female soldiers, because the patterns of response to physical and cognitive stressors are likely to be different in male and female populations. Male and female may not be the only grouping that should be considered. The potential for other relevant subgroups underlines the need for an ability to recruit substantial numbers of research subjects.
- Here and elsewhere in HRED, more systematic thinking about what constitutes a significant result would be valuable. There are two relevant senses of significance. The first is statistical significance; HRED researchers have been improving markedly in their statistical analyses of data. The second is the significance of the finding to the Army and/or scientific community. Sometimes (as in public health research), very small effects can be both statistically and practically significant. Other times (as occurs quite frequently in cognitive research), it is possible to achieve statistical significance without having much scientific or practical impact. When planning for research in the physical and cognitive performance area, it would be valuable to try to determine how large an effect would be needed to be interesting as either a basic or applied result.
- Some studies appear to terminate without discussion of further research avenues. Ideally, clever, successful, single experiments would give rise to more programmatic series of experiments.

Overall Quality of Research

The research of the physical and cognitive performance interaction research group is generally of high technical quality. To date, its scope has been somewhat limited, but its work may be viewed as preliminary to a successful, coherent program of research based on a developing vision and an excellent suite of deployed experimental platforms.

TRANSLATIONAL NEUROSCIENCE

Changes Since the Previous Review

Two changes are noteworthy. First, the program is now referred to as "Translational Neuroscience" rather than simply "Neuroscience." This is not mere jargon, but reflects a continuing refinement of HRED's goals in neuroscience. The goal of the translational neuroscience program at HRED is to integrate modern neuroscience with human factors, psychology, and engineering to enhance our understanding of soldier function and behavior in complex operational settings. Neuroscience is a vast field, and it has been imperative for HRED to concentrate its efforts on specific problems with potential Army relevance. The second notable change is the arrival of a senior scientist in neuroscience, who is also an ARL Fellow and whose background brings expertise in computational methods of signal analysis and neural modeling to HRED.

Accomplishments and Advancements

Over the past 2 years, translational neuroscience has made impressive strides at HRED. At the time of the previous review, the neuroscience group was in its promising infancy. Progress during the past 24 months, driven by strong leadership, provides a model for how a new group should be developed at a government research laboratory. During the past 5 years, ARL's neuroscience group has grown into the Department of Defense's (DoD's) largest internal nonmedical translational neuroscience research effort. A central task during this period has been to refine the neuroscience group's vision in a way that allows it to be Army-relevant while continuing its trajectory toward being a neuroscience laboratory on a par with strong university research programs.

To develop this Army-relevant, translational research capability, the group has concentrated its efforts on three internal research thrusts:

1. *Brain-computer interaction technologies.* What kinds of neurotechnologies have potential for broad DoD and civilian impact? Where is basic research required before translational research will be possible?
2. *Real-world neuroimaging.* What aspects of brain function can be usefully monitored outside of the laboratory setting? What are the technologies that are best adapted for this purpose? For example, would it be practical to record aspects of EEG from a soldier as he/she is driving on a mission? Would there be reliable biomarkers for stress or fatigue in the signals?
3. *Individual differences and neurocognitive performance.* Assessing the different capabilities of soldiers is important if those soldiers are to be optimally matched to tasks. What is the role of neuroscientific measures in assessing individual differences?

There are a palpable energy and enthusiasm among the strong mix of early-career (HRED has added more than 10 postdoctoral researchers) and mid-career scientists. They seem to work well together and seem quite well connected with academic neuroscience research labs. As noted, above, a senior researcher has now been added to this mix. Connections to the broader scientific community include partnerships with more than 20 national and international universities and companies through programs including the Cognition and Neuroergonomics Collaborative Technology Alliance (C&N CTA) and the Institute for Collaborative Biotechnology (ICB).

The neuroscience group is to be commended for its excellent publication record. While concerns about publication have been raised in the review of other HRED groups, the neuroscience group has maintained an excellent publication record that would be appropriate at a good academic lab. The group has published at least 20 journal articles and 10 technical reports over the past 2 years. Moreover, the group is publishing in strong, peer-reviewed journals.

An example of the type of work that defines this group is that on detection and classification of artifacts in EEG signals. Typical EEG research is performed in quiet rooms with stationary subjects who have been recruited into the EEG study. If one anticipates recording EEG from soldiers who are moving through a stressful environment and who have other things on their mind besides data quality, then those standard laboratory conditions will not apply. The recording environment introduces substantial artifacts into the EEG. The goal of this research is to detect and then acknowledge and/or remove the artifacts so that the underlying signals can be interpreted. Although not presented as a definitive solution to this very substantial problem, the group's work represents a good start on developing methods that would be sufficiently computationally simple to permit online use. The expertise of the new senior scientist in modeling EEG data should be valuable in projects of this sort.

Opportunities and Challenges

In the case of the translational neuroscience program, many of the opportunities and challenges consist of recommendations to do more of what the group is already doing. The translational neuroscience group should:

- Continue to build outside collaborative networks.
- Continue to grow internally, by adding expertise in computational modeling.
- Work to develop methods to remove as well as detect EEG artifacts.
- Incorporate neural source modeling (generator localization methods).
- Consider multiple measures of functional connectivity (e.g., phase coherence, phase lag, and Granger causality).

Other groups within HRED and elsewhere in the Army are doing relevant neuroscience. Improving communication with those groups would assist this group in its work. The translational neuroscience group seems to be the obvious locus of leadership for all HRED-relevant neuroscience research.

Although the arrival of the senior scientist has associated benefits, there is an implicit challenge to integrate a senior investigator into HRED in a manner that will forward his goals as well as the goals of the translational neuroscience group.

Overall Quality of Research

The translational neuroscience group's research is on a par with work in the university community. This work is at the level of a good, university neuroscience department. This is not to say that each project will transform either fundamental or applied neuroscience research. However, the group conducts high-quality neuroscience research that is routinely validated by its publication in good, peer-reviewed journals.

SOCIAL-COGNITIVE NETWORK SCIENCE

Changes Since the Previous Review

During the past 2 years, the social-cognitive network science group has worked to refine its vision. Concretely, this has involved a reduction from five to two thrusts in sociotechnical network operations and network-enabled cognition and an increased effort to align work with that of the ARL's network science program. Consequently, in 2012 this area was examined as part of a review of the ARL network science program, managed by the Computational and Information Sciences Directorate. Discussion of that review can be found in Chapter 2 of this report.

Accomplishments and Advancements

The social-cognitive network science group is working on a significant set of Army-relevant problems. Soldiers will be increasingly making decisions as part of a dispersed and distributed network. Optimizing performance in these settings is an important goal. Moreover, the way in which soldiers interact with networks is a topic that is clearly related to HRED's core interests in human-systems interactions. Connecting more closely with the broader network science endeavor can serve to keep everyone focused on the reality of the human element in most of these networks. There is a growing network aspect to the core HRED mission of human-systems integration. Some aspects of the research program have great promise. For example, the ability to study networks during field exercises is an opportunity that is unavailable to most other scientists, and the group has made some interesting progress with this approach.

Opportunities and Challenges

The following are central concerns with the work in this area:

1. The vision for the group remains a work in progress. Social-cognitive network science (like any of the HRED research thrusts) is a vast topic, and choices should be made about where to devote research effort and where to build staff expertise. HRED's answer to this concern has not crystallized for the social-cognitive network science area.
2. Concerns about the quality of the research in this area remain. Good work is being conducted, but too many presentations of the work suffer from problems such as:
   a. *Unclear goals.* Why was this specific problem deemed worth study? What is the theoretical framework?
   b. *Weak statistical analysis.* This comment is based on observations during the 2011 briefing. Social-cognitive work was not briefed to the panel in 2012, and so this report does not present evaluation of progress on statistical analysis that may have occurred.
   c. *Some studies seemed underpowered.* For example, sweeping conclusions cannot be based on comparisons of two teams that differ from each other on a variety of variables.
   d. *Many studies seem to be one-shot studies that do not build a cumulative body of knowledge.*

With these challenges come opportunities. The refinement of the vision of this group may make it clear that the group needs to bring in new expertise. Addressing the challenge of network science may require staff with backgrounds different from what is currently available in HRED. Some of this expertise can come from external partners, for example, via the CTA process. It will be important for this group to interact closely with the external researchers working on related issues.

Overall Quality of Research

The overall quality of the research in this area is uneven. There is some fine work but much work will have low impact, either because it is methodologically weak or because it does not appear to be part of a systematic program of work. Working more closely with the opportunity-driven researchers could assure that the work in social-cognitive networks is well grounded and relevant to field operations.

HUMAN-SYSTEMS INTEGRATION

Changes Since the Previous Review

Incorporation of the impact of social and cultural influences into human-systems integration is a new departure for this group.

Accomplishments and Advancements

HSI represents one of HRED's core competencies—providing the Army with assessments of how humans will work with new systems. The HSI program works to develop tools to model human performance and to evaluate performance tradeoffs that will occur when factors such as cost, size, weight, user interface, and other factors are changed. By using effective human performance modeling tools early in the acquisition process, potential performance problems can be identified and resolved sooner resulting in better-designed systems for the warfighter. This work requires a combination of knowledge of the Army materiel acquisition system and of the constraints on human performance. Ideally, if effective human performance modeling tools are used early in the acquisition process, potential performance problems can be identified at this early stage, and expensive, potentially dangerous problems can be avoided. HSI analysis might, for example, offer informed opinions about the numbers of crew required for a vehicle, or it might determine whether increasing the armor on that vehicle reduces the ability to navigate in an unacceptable manner.

Currently, three main research thrusts support the HSI group's evaluation mission:

1. *Development of tools*. The HSI goal in tool development is to develop or improve tools that allow early evaluation of human-systems integration. MANPRINT is the Army's programmatic implementation of HSI. MANPRINT is a set of human factors requirements intended to optimize total system performance, reduce life cycle costs, and minimize risk of soldier loss or injury by ensuring a systematic consideration of the impact of materiel design on soldiers throughout the system development process. IMPRINT is HRED's primary tool supporting MANPRINT. HRED describes IMPRINT as a dynamic, stochastic, discrete-event network modeling tool designed to help assess the interaction of warfighter and system performance throughout the system life cycle—from concept and design to field testing and system upgrades.
2. *Analysis methodology*. Early, rapid, cost-effective MANPRINT analyses are the traditional heart of the HRED HSI enterprise. HRED works to improve this process by increasing the coordination among the research, development, testing, and warfighter communities. The goals include ensuring that MANPRINT analyses occur early enough and that the results and recommendations arising from those analyses are tracked by the program manager in an effective manner.
3. *Socio-cultural behavioral modeling*. This third thrust area is relatively new for the HSI group. Research is aimed at understanding and modeling cognitive aspects of socio-cultural influences on soldier and/or commander decision making and communication. Models of decision making

and of communication should account for socio-cultural influences. The work is intended to produce guiding principles for presentation of socio-cultural information.

IMPRINT is the crown jewel of HRED's HSI efforts. It represents a mechanism for applying human factors science through modeling. Its continued development and use constitute a strength of the program. IMPRINT development can be monitored at the IMPRINT website (http://www.arl.army.mil/www/default.cfm?page=445). There may be a gap in IMPRINT in the form of inadequate coverage of cognition and perception. If so, given the nature of today's Army tasks, this is a critical gap.

The use of HRED's HSI expertise can be seen in work such as that on progressive insertion of human figure modeling into acquisition. Good human figure modeling tools are of obvious use in the design of systems that humans are going to use. However, models for Army use are not going to be simple, off-the-shelf civilian models. Civilians do not need to enter and exit vehicles wearing body armor, hydration packs, extreme cold weather gear, chemical protective equipment, or other gear. Designing for the classic 5th and 95th percentiles may be misleading if the population is not unimodal. HRED's effort to improve models of soldiers is a cost-effective way to improve the design of systems during the acquisition process. The research has provided effective comparisons between model-predicted outcomes and actual outcomes.

Opportunities and Challenges

Research in the human-systems integration area is uneven. Much of the work, such as that described above, is solid—part of an important effort to insert human factors into the acquisition process. Other work is less impressive. For example, the presentation of the work on multiple simultaneous task experiments revealed the types of problems that afflict some of the weaker work in this area. This project was designed without adequate attention to the existing literature on, for example, effect of cognitive load or dual task interference. An ad hoc task design was neither realistic enough to be as compelling as applied research nor basic enough to be likely to generate generalizable basic research findings. The statistical analysis was weak, and, in the absence of theory, it was unclear how these results could be used to predict or estimate performance in any other task. The intent here is not to single out one study, but to point out recurring problems with the design, analysis, and interpretation of studies.

The new social-cultural thrust faces another set of challenges. The study of emerging work in social, cultural, and behavioral modeling is interesting and quite appealing but elicits serious programmatic reservations. The problem area is vast. To make any serious progress in this area, HRED would need to devote significant resources to the endeavor. It would need to assign multiple investigators to the problem, some of whom would have to come from outside HRED because the requisite expertise does not exist in house. That decision support is a good endpoint for the study has not been convincingly justified. HSI has dipped a toe into these interesting waters. It either needs to plunge in with a serious investment of resources or leave the topic to others.

As previously noted, it is possible that cognitive and perceptual variables are treated too lightly in IMPRINT. If this is the case, then it would be valuable to address this deficiency. Doing so represents an opportunity for HRED to bring together investigators who have been developing physical and perceptual/cognitive models in other domains for the purpose of providing a more robust human model of soldiers than is available in the existing IMPRINT model.

Overall Quality of Research

As indicated above, the picture in this area is uneven. Much of the work is very solid, high-quality human factors work, employing a range of tools, including IMPRINT. Other work is deficient in conception and/or execution.

OPPORTUNITY-DRIVEN RESEARCH

Changes Since the Previous Review

Since the previous review, opportunity-driven research (ODR) has been elevated to the status of a research thrust. This development is very positive. The value of careful management and nurturing of this pathway cannot be overstated. ODR can be a central aspect of HRED's mission to enhance the ability of soldiers to work with Army systems, to neutralize threats to those soldiers, and to increase affordability. Moreover, successful response to problems that present in the ODR context can enhance the credibility and perceived relevance of HRED. This, in turn, seems likely to build external advocacy for the Army investment in HRED research.

Accomplishments and Advancements

HRED frames ODR as having a close relationship to the HSI/MANPRINT work, described above. Although the HSI/MANPRINT effort is targeted toward intervening early in the acquisition process, MANPRINT issues can be identified during usability testing or through reports received from the field. When there is no previous research to provide a solution to a problem, an opportunity arises for research. This problem-solving approach to MANPRINT issues is an excellent opportunity for HRED to demonstrate its unique talents. This sort of problem-inspired research and development seems central to the HRED niche. ODR enhances the credibility of the science program and should be an incubator for further research.

ODR also fulfills a soldiers' advocacy role when issues identified in the field rise to the level of science rather than yield mere complaints or ad hoc fixes. Much of the ODR work is conducted in HRED's 20 field elements, although it does occur elsewhere in HRED, which makes sense because the field elements are HRED's forward positions. The field-element-based work can be the first step in work that would be later developed at Aberdeen.

By its nature, ODR is not systematic in the way that a program such as translational neuroscience would be. Consequently, it is not possible to comment on ODR as a whole based on reports of a few programs. However, in general, the projects that were presented bring good human factors research to interesting problems of obvious relevance. There is potential here that is not being fully exploited.

Opportunities and Challenges

Current HRED doctrine is that the goals of ODR are to fill critical knowledge gaps and to support the solution of real-world problems for a specific group of end users. As such, it does not consist of large lines of research geared toward creating a new generalizable body of knowledge. That ODR does not involve large lines of research is unassailable. But to conclude that ODR does not lead to generalizable knowledge limits its potential impact. In some cases, ODR research could lead to important, larger lines of research with implications for the Army as well as potential to generate good fundamental research.

An example is the work on investigating takeoff and landing tasks in a brown-out degraded visual environment. This was a very fine presentation on an important problem. Obviously, it is difficult for pilots to land a helicopter if they cannot see. Relatively abstract symbology about altitude and other factors related to landing is useful, but symbology that exploits pictorial depth cues can enhance situation awareness and improve performance. Clearly, if this work prevented one helicopter crash it would have been more than worth the investment. However, there would be no reason to stop at this point. The specific symbology made use of some depth cues in one configuration. This was a good first try, but could the display be improved further? HRED has a sensory performance group that should have the expertise to ask the more general questions about displays of this sort. Here is an example of a problem that could be brought from the field element into HRED Aberdeen for more systematic study. Improvements could be suggested and then transitioned back to the field element for testing.

As noted elsewhere within HRED, many projects are often done in response to some specific customer request and terminate without discussion of further research avenues. In some cases, this ending may be appropriate. However, in other cases, an opportunity may be missed. The specific request may require only a quick study and a good-enough solution. HRED scientific leadership should monitor ODR projects for opportunities to bring research from the field elements to Aberdeen, with the goal of producing better than good-enough solutions to the specific problem. Such solutions might have impact beyond the originating problem.

The project on enhanced operator perception through three-dimensional (3D) vision and haptic feedback raises similar issues. In this case, the question was whether a robot arm could be manipulated more effectively if the operator had better 3D input and haptic feedback. The answer from the solidly executed study was that these enhancements were helpful. Again, a program of research might yield stronger insights on this topic. In this case, it seems likely that best progress would be made in collaboration with robotics partners beyond HRED, whose role is on the evaluation side of the equation. Other ARL partners would be a possibility, as would industry or academic partnerships arranged through the Army Research Office or a funding vehicle like a CTA. This particular study is highlighted only as an example; the ARLTAB is not advocating for a massive project on robotic arms. HRED leadership should view ODR projects as seeds of larger research projects.

Not all ODR projects need to grow into research programs. An example is the work on evaluating training using the Second Life video game platform to augment mindfulness training for soldiers. Although the topic has value, this project did not include plans for follow-on research and did not fall obviously under the HRED umbrella. If it were to be pursued, then the Simulation & Training Technology Center would be a more obvious home for more extensive study.

Overall Quality of Research

The ODR work that was presented during the reviews was of generally high quality. This work nicely illustrates the ability of HRED to serve as a problem-solving resource for a wide range of Army problems.

SIMULATION AND TRAINING TECHNOLOGY

Changes Since the Previous Review

The previous review occurred as the Simulation and Training Technology Center (STTC) was being integrated into HRED. Therefore, this is the ARLTAB's first review of the STTC.

Accomplishments and Advancements

The STTC is a large group of researchers and developers based in Orlando, Florida. The group became part of HRED within the past few years. The STTC's efforts are directed at advancing the Army's simulation-based capabilities in training, experimentation, analysis, and operational Army needs. Simulation is seen as a cost-effective response to these needs. The pace and diversity of U.S. Army missions require a rapid, responsive training capability. Moreover, in a variety of situations (e.g., adaptive tutoring), training by the use of artificially intelligent agents has been shown to produce better results than has more conventional training.

The STTC designs specific simulations (e.g., simulations of battlefield medical situations), as well as general tools for simulation (e.g., the Generalized Intelligent Framework for Tutoring framework) to make it possible for others to rapidly create new training modules for new content areas.

The STTC has the program management for a University Affiliated Research Center, the Institute for Creative Technology (ICT) at the University of Southern California. The ICT receives 100 percent of the STTC basic research (6.1) funding. The ICT is reviewed by a separate Army assessment board. The ARLTAB was not asked to review the ICT, which limits the scope of the evaluation of STTC presented in this report.

The STTC includes five programs:

1. The purpose of the *Adaptive and Intelligent Training Technologies Program* is to design, develop, apply, and assess artificially intelligent agent technologies (e.g., adaptive tutoring and virtual human tools and methods) to enhance training effectiveness and reduce costs. An important emphasis is on developing tools that allow others (i.e., researchers, instructional designers, training developers, and trainers) to quickly author new training modules so that artificial tutors can be created for different training needs as they arise. The conscious effort to develop a broadly applicable framework has produced the STTC's generalized intelligent framework of tutoring (GIFT), which is a sensible and useful tool for the development and evaluation of intelligent training systems.
2. The *Synthetic Environments Program* develops improvements in synthetic environment modeling, with a particular emphasis on dynamic effects such as changes in weather and lighting.
3. The *Immersive Learning Program* investigates the issues raised when these synthetic environments are used as learning environments. In addition to studying what makes for a realistic, compelling environment, the program is attempting to address whether or not immersive environments actually promote learning. To the extent that the immersive situation provides an effective version of time-on-task training, it would be expected to lead to better learning outcomes (e.g., skills acquisition, retention, and transfer to the field).
4. Obviously, the above work is related to the *Training Applications Program*, which creates and tests specific examples of training applications software programs. The training applications program focuses on domain-specific research such as medical training, dismounted soldier training, or ground platform training.
5. The *Advanced Distributed Simulation Program* focuses on conducting research and developing technology to facilitate local and geographically distributed interactions between models and simulations.

These programs constitute a generally impressive program of work, although comparatively few program details were presented. The panel has not seen the facilities nor met many of the staff.

STTC staff has made impressive efforts to link STTC work to broader work in the field.

Opportunities and Challenges

The main challenge lies in the integration of the STTC into HRED in a manner that is useful to both. Any cognitive psychologists associated with the STTC in Orlando could help to facilitate an increased focus on human factors in training.

Some of the work will likely benefit from a closer link with HRED. The work on real-time monitoring of ECG and GSR signals during computer-based training is a natural point of collaboration with the translational neuroscience group. The current use of galvanic skin response and EEG to measure student arousal seems intrusive and not particularly practical for complex, dynamic training situations. The relationship between EEG and GSR signals and engagement was not obvious, and the generalizability of findings across different contexts and tasks was not clear.

Overall Quality of Research

The overall quality of the work reviewed is good.

4

Sensors and Electron Devices Directorate

INTRODUCTION

The Panel on Sensors and Electron Devices reviewed the research activity of the Sensors and Electron Devices Directorate (SEDD). The panel met at the Army Research Laboratory (ARL) facility in Adelphi, Maryland, on May 10-12, 2011, and May 22-24, 2012. During those two meetings, the panel reviewed research portfolios in all four SEDD divisions: Electro-Optics and Photonics, Energy and Power, Electronics and Radio Frequency, and Signal and Image Processing.

The review focused on both internal research projects and collaborative activities. SEDD is a lead directorate in the new Multiscale Multidisciplinary Modeling of Electronic Materials (MSME) Collaborative Research Alliance (CRA) and is a participant in multiple Collaborative Technology Alliance (CTA) programs focused on robotics, network sciences, cognition and neuroergonomics, and micro autonomous systems and technology. Several of its research centers and institutes focus on flexible displays, fuel processing, biotechnology, nanoscience, and microelectronics manufacturing. SEDD also participates in the International Technology Alliance (ITA) program in network science.

CHANGES SINCE THE PREVIOUS REVIEW

One of the most notable changes in SEDD is the evolution of the technical and scientific staff, which now numbers 332. SEDD employs a remarkable group of enthusiastic and talented people who are excited about the work they are doing and the environment in which they are doing it. In this cycle SEDD added talented staff in areas such as bioscience, intelligence analysis, and quantum information processing despite the highly competitive job market in these areas. Moreover, SEDD is to be congratulated for its successful recruitment of highly talented and impressive women and minorities among the new staff members. ARL's programs supporting student interns, thesis projects, and postdoctoral fellows also are praiseworthy.

Improvement is also evidenced by increases in the numbers of publications, patents, and awards for scientific excellence. Refereed publications in 2011 increased by 36 percent and patent awards by 24 percent when compared to the average of the previous 4 years. Awards and recognition include 2 best paper awards, 2 scientific achievement awards from professional societies, and 11 Secretary of the Army Research and Development Awards, as well as election of a senior staff member as a fellow of the American Physical Society.

SEDD has made a significant commitment to maintaining state-of-the-art microelectronics fabrication facilities in-house through the Specialty Electronics, Materials, and Sensors (SEMACS) program. To this end, it invested $13.7 million over the past 2 years in new and updated cleanroom facilities and in other new facilities, including the Microsystem Indoor Testing Grounds developed for validation of autonomous microsystem devices under the Micro Autonomous Systems and Technology (MAST) program. The facilities provide essential support for Army-niche research and engineering activities in such areas as electro-optics and photonics, electronics and radio frequency devices, sensors, and power and energy.

During the current review period, SEDD developed new relationships with industry and other Department of Defense (DoD) agencies and strengthened existing partnerships. Standouts among these relationships are the Flexible Display Center (involving more than 30 partners from industry, academia, and government, sharing a common goal of developing full color, video rate, flexible display technology for commercial and military applications); new partnerships in compound semiconductor devices with TriQuint Semiconductor and Cree, Inc., who perform manufacturing of compound semiconductor devices; and the MEMS (microelectromechanical systems) Exchange, which does device design and fabrication through a network of commercial, academic, and government foundries.

SEDD has taken a leadership role in the ambitious new MSME CRA, which focuses on the development of a numerical toolset that can be used to build models that span many physics domains on a number of specific problems. With the breadth of knowledge at SEDD, and with the strong ties to the academic research community, SEDD and its partners are positioned to have a high impact in this area.

ACCOMPLISHMENTS AND ADVANCEMENTS

This section is organized by each of the SEDD divisions—Energy and Power, Electronics and Radio Frequency, Electro-Optics and Photonics, and Signal and Image Processing—and highlights some of the most significant accomplishments for projects within each.

Energy and Power Division

During this cycle, significant and encouraging advances in the division's research vision and outlook reflect the evolution of the energy and power field in general, as well as the challenges facing the Army and ARL in particular. Division leadership recognizes that now is a great time to take advantage of the host of opportunities being offered to energy and power. Discussed below are some of the most notable recent accomplishments.

Micropower

The notion of a fully integrated (electronics and passives), single-chip power supply is an elusive goal in the power electronics business. SEDD researchers have made important progress, producing results that are among the best in the field. The first experimental results show Q values for air core inductors greater than the state of the art. The magnetics are tunable to a remarkable degree, presenting the possibility for improved voltage gain. This is one of the division's most remarkable successes, with potential application to a fundamental Army need for miniature size and weight of power supplies.

Passives

SEDD's work on high-performance and high-value passives has significantly advanced the state of the art. In particular, SEDD has demonstrated a miniature on-chip inductor fabricated by thick electroplating with high quality factors and high inductance density, which is the enabling technology for microscale power conversion technology.

Smart Grid Technology

SEDD researchers have begun building the tools and components necessary to implement smart grid technology in the tactical Army environment. They are building a hardware test and experimentation facility to model the tactical command post's integrated electrical power environment. Equipment includes a real-time digital simulator (RTDS)—cutting-edge equipment in the public electric utility industry—to enhance flexibility. SEDD has partnered with other agencies, notably the Department of Energy, for specific proposals, which aligns with the anticipated technology handoff to the U.S. Army Communications-Electronics Command (CECOM), which already has initiatives to do the tactical integration and networking.

Related work in intelligent power management, energy harvesting, and micro-grid implementations has shown results that will enable soldiers to increase their reliance on electrical/electronic devices by reducing the weight loads on their shoulders.

Fuel Cells

Significant progress has been made in the area of fuel cells. In particular, a field test of an M100 direct methanol fuel cell system demonstrated that fuel cells can provide a high energy density, that will reduce logistics and provide both cost savings and weight savings. The new MSME CRA on electronic materials and electrochemical systems will substantially impact this program, and it will leverage computational expertise in academia with the experimental expertise at ARL.

Batteries

SEDD researchers have developed an additive to the electrolyte of lithium-ion batteries that modifies the solid electrolyte interface/interphase (SEI) layer chemistry and improves its stability, which has helped SEDD demonstrate the excellent voltage stability of Li-ion cells. This effort also involved computational work that proposed a reaction mechanism for the formation of the SEI layer. This is a good example of joint computational and experimental work.

Work on the Li-Air battery also produced promising results toward developing a low-cost non-precious-metal-based catalyst: the developed catalyst lifted the discharge voltage by at least 0.2 V, improved discharge rate capability, and reduced cell reaction activation energy. Work on Li-La-Zr-oxide doped with Al to stabilize the cubic phase as solid electrolyte for lithium-based batteries also shows significant promise.

Silicon Carbide (SiC) Electron Devices

SEDD continues to be the leader in research in this field. SEDD has demonstrated a resonant 40 kV, toward 120 kV, DC/DC converter that uses SiC's advantages of high voltage and perhaps higher temperature capabilities to meet SWaP (size, weight, and power) requirements. SEDD also has identified important packaging issues in the initial prototype.

In other power electronics efforts, the implementation of SiC devices from large-diameter wafers for DC through pulsed applications is a significant achievement. Wide band gap semiconductors, particularly SiC, is a niche that ARL has filled well. ARL continues to be the leader in research in this field. Manufacturers continue to go to ARL for technical expertise. Similar efforts to find solutions in gallium nitride (GaN) and GaN-based materials have also seen progress.

Extreme Energy

In this transformational new project SEDD scientists collaborate with colleagues in biological sciences, physics, chemistry, and engineering to develop a technology for synthetic photosynthesis. With its interdisciplinary workforce and well-equipped laboratories, SEDD is well poised to succeed in this highly speculative enterprise.

Automatic Reconfiguration of Power

SEDD has begun to look at technologies that address power reconfiguration challenges. Specifically, there is a practical requirement to reconfigure power after battle damage and an economic advantage to reconfiguring equipment after it is deployed. SEDD recognizes this need and is investigating ways to address these issues in the context of Army environments and equipment.

Electronics and Radio Frequency Division

The portfolio of the Radio Frequency and Electronics (RF&E) Division ranges from fundamental electronic devices and sensor research to high-performance radar systems. This section highlights the recent accomplishment from this largest of the SEDD divisions.

High Speed Digital to Analog Conversion

SEDD researchers have developed advanced design techniques for ultra wideband digital-to-analog converters (DACs). In this cycle, they have designed and simulated an 8 bit, 32 GS/s DAC, which will be an essential component in future software-defined radio technology.

MEMS

SEDD maintains a thriving MEMS capability. The program's applications range from RF components, to antennas, to small unmanned aerial vehicles. Accomplishments include fundamental materials development in graphene and piezoelectric and near-term devices such as the traumatic brain injury sensor. The latter is an outstanding achievement that will have an immediate impact on soldiers in the field.

Ultra-Wideband (UWB) Radar Processing

Researchers have developed a clever algorithm based on single-value decomposition that reduces the amount and intensity of clutter returns while maintaining target returns. In addition, a second algorithm has been implemented that uses compressive sensing to eliminate the degrading effects of interfering signals. This may be the first successful use of compressive sensing for this application.

Electro-Optics and Photonics Division

This section highlights several projects from the Electro-Optics and Photonics Division (EO&P). Although biological science research is also housed in EO&P, SEDD undertook in 2012 a separate review focused exclusively on ARL research related to biological science. This section focuses solely on EO&P projects.

Quantum Dots

Quantum dots have been known to enhance the light absorption through multiple energy levels in photo sensors. They can also extend the absorption edge into the infrared range of the light spectrum. SEDD researchers are investigating the effects of quantum dots with build-in charge on the solar cell harvesting and recombination process. SEDD researchers have achieved an increase in the power efficiency of an InAs/GaAs quantum dot solar cell from 9 percent to 14 percent.

High-Performance IR Detectors

Resonator quantum-well infrared photodetectors (QWIPs) continue be a crown-jewel achievement of SEDD. SEDD has been able to incorporate advanced optical concepts into the detector design. SEDD has developed a better understanding of new near-field optical phenomena that has enabled design of a QWIP structure with quantum efficiencies of nearly 70 percent, doubling the previous record of 35 percent for corrugated QWIPs.

SEDD is also investigating advances in infrared detectors based on HgCdTe. Specifically, the approach is to replace tellurium (Te) in the existing HgCdTe by selenium (Se). This is work in progress, but preliminary results indicate that some improvements have been obtained.

Optoelectronic Oscillator

EO&P research aimed at creating an extremely low noise oscillator has produced outstanding results. SEDD researchers have found a way to build a 10-GHz optoelectronic oscillator that matches the specifications of commercial low-noise oscillators up to 10 kHz away from the carrier at an order-of-magnitude lower cost and competitive size and weight specifications. This accomplishment will be of great interest to the radar community for the detection of small targets in clutter.

Cold Atom Optics

The underlying notion behind cold atom optics is that everything in conventional electronics, from wires to batteries to transistors and diodes, can be replaced by atom analogs. The project's objective is to explore the possibility of using atom spin instead of electron spin to make novel spintronic devices.

Specifically, the spin polarization of atoms trapped near the surface of a microfabricated semiconductor chip will be controlled, eventually to transport the atom spin along the surface of the chip. Ultimately, this could lead to an atom-based spin transistor. Spin-polarized atoms will open up more possibilities for advanced technology in computation, signal processing, and sensing than are available with spin polarized electrons. Coherent control and entanglement of neutral atom qubits confined on an atom chip by means of an optical frequency comb can potentially replace the well-known trapped ions. This technology relies on a laser to cool atoms to very low temperatures. The temperature achieved in a laboratory demonstration of this technology was a remarkable 40×10^{-6} K.

Ultraviolet Lasers

SEDD, in collaboration with the University of Central Florida, has demonstrated high-quality AlGaN/Ga:MgZnO grown on sapphire, which is extremely important for implementing high-efficiency near-ultraviolet lasers.

Signal and Image Processing Division

A large part of the signal and image processing effort within SEDD is collaborative work with CISD as part of the Network Science Program. This research is discussed in Chapter 2. This section highlights additional projects within the Signal and Image Processing (S&IP) Division.

Image Processing for Facial Recognition

SEDD researchers have demonstrated a capability to enhance the performance of low-resolution security cameras so that the images can be used to improve the verification rate of automatic facial recognition algorithms. This important result will enable more effective use of security cameras to protect guarded facilities and will allow law enforcement to make far better use of security cameras to address serious threats.

A parallel effort has focused on devising machine recognition methods for detecting humans in stationary scenes. This collaborative effort draws on SEDD's machine recognition skills and super-resolution technology and CISD's social-cognitive work. Because human experts have almost always outperformed every other approach, the long-term promise of this technique is enormous. This work demonstrated the ability of recognition algorithms to identify humans in challenging scenes, such as those with partially obscured targets.

OPPORTUNITIES AND CHALLENGES

SEDD faces the challenge of building an innovation ecology that generates novel technology to enable the warfighter. In doing so, it needs to create an organic and bottom-up environment for its researchers to innovate, yet this environment should be constrained to insure that the needed technological innovation is there to support the warfighter. Balancing the need to constrain an open technology innovation environment is not an easy challenge. SEDD leadership is aware of the delicate balance that they need to have in place to catalyze innovation and the need to actively, but not intrusively, intervene to keep the ecology in synchrony with the SEDD mission.

The rotating SEDD directorship is not healthy for the organization. Talented division chiefs have served as director for 4-month intervals, which is not sufficient time to allow for (1) sustained and consistent assessment of how the customers' needs match with the current research and development activities; (2) alignment of funding and personnel resources in the light of shifting missions and budgets; and (3) tweaking of the constraints to adjust the flow of innovation. A permanent director is needed for SEDD.

The recently initiated multiscale modeling CRAs are ambitious yet necessary efforts. ARL is to be commended for establishing the two CRAs, the MSME, and the Materials in Extreme Dynamic Environments (MEDE). Under the MSME, a multi-university team is charged with doing the fundamental multiscale modeling on lithium batteries, fuel cells, electronic materials, and devices. ARL will perform the supporting experimental work. Given the substantial issues concerning the premise of bridging the scales in multiscale modeling, many challenges (and thus opportunities) exist in demonstrating verifiable success in the 5-year timeframe. Close interactions of ARL personnel with the multi-university team selected will be critical to ensure a focused and productive group effort. A potential challenge is to make sure that the multi-university team, with principal investigators having different (not necessarily overlapping) skills, remains focused and delivers value to ARL.

Work on the MAST program has been impressive. However, supplying ample power to autonomous devices is a challenging problem, and it is not clear whether a solution is in sight. Work to design a fully integrated MAST with power onboard may be needed and may lead to a system design that is achievable in a realistic timeframe.

ARL has outstanding fabrication facilities. Many projects that use the facilities also require microstructural characterization at the nanometer scale. SEDD, and more generally ARL, should make every effort to ensure that characterization facilities, such as high-resolution and analytical electron microscopy, are readily available. Moreover, in-house characterization capabilities will have to keep pace with the quality of the fabrication equipment.

OVERALL TECHNICAL QUALITY OF THE WORK

The quality of research being conducted within SEDD is excellent. This is reflected in the numerous awards received by the SEDD scientific staff, by the number of refereed publications authored and coauthored, and by the number of presentations given at professional society meetings. By these metrics, ARL compares favorably with other private and university-based research institutions. There is no doubt that SEDD is conducting world-class research.

SEDD researchers are efficient at leveraging ongoing research that appears in the technical literature, at technical conferences, and through contacts with active researchers. They have the enormous task of understanding the needs of an astonishingly broad range of Army applications and then providing innovative solutions. From a technical viewpoint, they are performing this mission quite well.

It is impressive that in this context SEDD endeavors to predict Army needs that are one to two decades out. For example, extreme energy science has been identified as a strategic initiative for the laboratory. In response, SEDD staff has successfully combined in-house capabilities and external collaborations to quickly lead the scientific community working in this nascent field.

The ARL/SEDD policy of involving undergraduates, graduates, and postdoctoral fellows as part of in-house research efforts is not only a wonderful vehicle to educate the nation's workforce but also a great coupling with educational institutions that might or might not have formal research contracts with ARL. This practice is beneficial to all involved, and its continual implementation should be encouraged.

5

Survivability/Lethality Analysis Directorate

INTRODUCTION

The Survivability/Lethality Analysis Directorate (SLAD) was reviewed by the Panel on Survivability and Lethality Analysis during January 25-26, 2011, and August 3-5, 2011, at the White Sands Missile Range in New Mexico, and during May 29-June 1, 2012, at the Aberdeen Proving Ground, Maryland.

SLAD describes itself as follows:

> SLAD is the premier source of expertise in survivability, lethality, and vulnerability (SLV) assessments, for senior leaders, developers, and evaluators, helping to ensure that U.S. personnel and equipment survive and function effectively in hostile circumstances. SLAD conducts analytical investigations; modeling and simulations; and laboratory and field experiments to provide its analyses as well as technical advice, and it strives to be a subject-matter expert on survivability and lethality matters to program executive officers (PEOs), program managers (PMs), users, testers, the Army's independent evaluator, and other customers. SLAD produces products intended to reveal critical survivability and lethality issues for Army milestone decisions. In order to best serve the Army's analytical needs, SLAD leverages both research and engineering conducted by other Army research, development, and engineering centers as well as other services, and SLAD's work is leveraged by others.[1]

SLAD consists of two divisions: the Information and Electronic Protection Division and the Ballistics and Nuclear, Biological, and Chemical Division. The SLAD portfolio consists of many small test and evaluation (T&E) programs directly related to assessment of specific Army components, with several larger programs aimed at developing tools necessary to enhance test and evaluation efforts and broaden the scope of SLAD's contribution to the life-cycle assessment of Army systems. During its 2-year project review cycle, the panel reviews a representative portfolio of SLAD work at a level sufficient to assess

[1]Army Research Laboratory website, http://www.arl.army.mil/www/default.cfm?page=33, accessed February 1, 2013.

in depth the technical quality of its work. This chapter focuses on the programs for which adequate depth of detail was provided and where strong program continuity of effort was demonstrated. It also discusses a set of new and mature programs, in order to provide the context of the full spectra of work performed within SLAD.

CHANGES SINCE THE PREVIOUS REVIEW

Toward the beginning of this review cycle, the criteria applied to the assessment were amended in recognition of the programmatic emphases within SLAD on applied work such as test and evaluation, analysis, and tool development to enhance analysis capabilities. The previous criterion that emphasized the quality of scientific research was therefore replaced with a criterion that emphasizes the quality of the analytical work and of the efforts to support, extend, verify, and validate analytical tools. *Joint Vision 2020* is the Department of Defense (DoD) vision that defines how the various elements of the DoD, including the Army forces, will operate in global conflicts as a single, integrated warfighting entity.[2] Coupled with this vision is recognition of the constantly evolving threat that requires U.S. military forces to adapt and respond more rapidly with modified tactics, technologies, and or equipment than traditional DoD doctrinal requirements and acquisition processes provide for. This vision defines the need for the development of analysis tools that focus on quickly getting the best equipment to soldiers in the field. SLAD has demonstrated an exemplary capability for rapid and innovative response to solving problems with deployed or soon-to-be-deployed equipment. SLAD extends this rapid, innovative response to include analyses that address the survivability and lethality of integrated systems of equipment and humans, and it creates the partnerships that will enable it to impact system development in the conceptual planning stages.

SLAD increasingly recognizes the need to take advantage of opportunities to do the following:

- Extend attention to survivability and lethality of electronic weapons; electromagnetic pulse devices; directed energy; nuclear, chemical-biological, biomedical, and information warfare; and other weapon systems;
- Upgrade SLAD capabilities in wide-area security; and
- Proactively establish SLAD's value at an earlier stage in the development cycle of Army materiel programs.

Following the demise of the Future Combat System (FCS), the emphasis of Army acquisition shifted to ongoing campaigns in Iraq and Afghanistan. Opportunities for heavily armored mounted warfare dwindled, and the challenges confronting infantry and other soldiers mounted in lightly armored vehicles or on foot were accorded the highest priority. While continuing development of its combined arms model, System-of-Systems Survivability Simulation (S4), and other research tools, SLAD also extended its research and development focus beyond the survivability aspects of equipment to that of the soldier. This has resulted in improved knowledge and capabilities. SLAD has enlarged its initiatives in human vulnerability assessment with multiple programs and expanding collaborations. The human availability technique methodology to analyze the combined performance of materiel and humans is a new initiative that examines the degradation of the performance of the whole materiel system while operating under combat conditions. This effort is a good example of increased collaboration between SLAD and the

[2]Department of Defense. *Joint Vision 2020*. Washington, D.C.: U.S. Government Printing Office. May 3, 2000. Available at http://www.fs.fed.us/fire/doctrine/genesis_and_evolution/source_materials/joint_vision_2020.pdf (accessed October 3, 2012).

Human Research and Engineering Directorate (HRED) to expand the capabilities of SLAD's assessment tools. A timely new program to develop a metric to predict mild traumatic brain injury represents a change from SLAD's past focus on materiel. These two programs, combined with many others focused on assessing behind-armor, blunt trauma, and injury due to under-body blast, demonstrate the SLAD commitment to ensuring soldier survivability.

ACCOMPLISHMENTS AND ADVANCEMENTS

Collaborations in all SLAD program areas continue to improve, and the contribution to quality of work and dissemination of results was evident. Collaborations in SLAD's core competency areas are moving toward excellence, and their continued expansion is encouraged.

The collaborative efforts between SLAD and HRED build on the expertise of each organization to rapidly develop a new anthropomorphic test device (WIAMan) for assessing injury from underbody blast (UBB). The program is well designed and has already provided excellent insights. One challenge is to ensure that the characteristics of the device make it applicable to females as well as males. Collaborations with academic partners for the same UBB injury assessment program area identified a work-around for critical issues involving limited distribution of live-fire test data. SLAD researchers have partnered with academia to get better data by using cadavers instead of anthropomorphic test devices (ATDs). By comparing the relative response between ATDs and cadavers, SLAD now has good data obtained under controlled conditions against which to benchmark WIAMan. This is an excellent example of the use of collaboration with organizations rich in experience in synergistic technology areas, leveraging SLAD expertise to expand their capabilities.

The program to develop a multispectral method for evaluation of an electro-optical/infrared sensor is another example of good collaborations with various government agencies. For example, the U.S. Army Aviation and Missile Research, Development, and Engineering Center (AMRDEC) is responsible for the battlefield integration of the selected sensor, and SLAD provides field measurements and multispectral sensor results. It was found that no single wavelength would provide sufficient detection of all possible threats, and so a multispectral method was developed. Collaborations seem well aimed to take advantage of other groups' expertise to reduce the size of the multispectral sensor and to collect additional data for developing empirical models. This program demonstrates a good mix of empirical work and analysis.

SLAD has partnered with the Computational and Information Sciences Directorate (CISD) to expand the capabilities of its S4 simulation tool to include the Wireless Emulation Laboratory model developed by CISD to emulate a mobile, ad hoc network. This collaboration could enable S4 to include real-time modeling of mobile network systems. This program is still quite young, and few results are available, but SLAD has the beginnings of a sound project plan and is partnering and collaborating with other services (Naval Research Laboratory, Air Force Research Laboratory, and Air Force Office of Scientific Research).

In several programs, innovation resulted in significant contributions to the advancement of methodology and the development of standards.

Historically, operational data have displayed a wide variability that is believed to result from variability of the experience of the data collectors. Live-fire experiments were designed to provide results that would coincide with operational information, thereby providing a controlled experimental basis for more consistent interpretation of operational data. By reinterpreting the operational data based on live-fire results, a more accurate estimate of vulnerability was obtained. These efforts demonstrate the excellent use of analysis to inform interpretation of empirical data, which can then be used to inform model analysis. This method should lead to better predictions of vehicle performance after underbody explosions.

An innovative new program takes advantage of available smart phone technology to expedite collection of more accurate field data in a consistent manner, resulting in increased data fidelity for analysis. This project is not likely to increase equipment in the field, because it is just a software addition to existing devices. The challenge lies in addressing the issue of data classification.

A new program to develop a metric to predict mild traumatic brain injury resulted in design of a surrogate sensor system that mimics the effect on the human brain of blast and blunt trauma. The program also proposed instrumentation for measuring the severity of the injury received. A sensor was designed and is now being developed. The next step is finding a means to allow the sensor to be more specific to the level of brain trauma. A process to be used in the field for rapid determination of the level of trauma has been proposed as well. This would allow the start of intervention that should reduce the long-term trauma effects. The approach is novel and thorough, and involves extensive collaboration with relevant forums.

Digital radio frequency monitoring is an innovative approach to expand utilization of a promising technology to analyze waveforms other than radar. The work is highly relevant, and the program has good engagement with the tactical community.

The Integrated Network Vulnerability Assessment Discovery Exploitation (INVA/DE) is an excellent tool for detecting the vulnerability of computer network operations. Further collaborations with CISD to pit SLAD's INVA/DE against CISD's Interrogator will test the robustness of each program. This impressive program seems very well tied into the intelligence community.

A core competence of SLAD is that of assessing the lethality and vulnerability of various weapon systems proposed for combat vehicles and the vulnerability of different design variants or system alternatives. The lethality and vulnerability analysis for the Ground Combat Vehicle program is an example of the tasks that SLAD routinely and rapidly accomplish for program managers. Its ballistic analysis of a lightweight vehicle-protection system is another excellent example of the strength of SLAD's core competence that demonstrates the value of SLAD analysis and testing to support Army acquisitions; SLAD involvement should be inserted early into the acquisition process when requirements are first being determined. The active protection systems (APS) program shows that SLAD has contributed to an area of work from which it had been largely excluded by Army program management and organizational assignments. SLAD personnel provided new results on the vulnerability of light-armored vehicles to the residual threat of rocket-propelled grenades, which can have a tremendous impact on future design and fielding of better APS systems.

SLAD is home to several state-of-the-art test and analysis facilities. At the Electromagnetic Vulnerability Assessment Facility (EMVAF) at White Sands, experiments are conducted to characterize the effects of electromagnetic signatures on vehicles and other equipment performance. SLAD has been working in this area for a long time, and the SLAD researchers are recognized as experts in the field by the technical community. This facility supports a large cross-section of collaborators within the federal government. SLAD recently made a good decision to hire a global positioning system (GPS) expert to develop programs to study GPS jamming. The group working at the EMVAF consistently does good work and proactively pursues appropriate partnerships and collaborations.

OPPORTUNITIES AND CHALLENGES

SLAD still suffers from an aging workforce and faces challenges in building and maintaining a strong, fresh, knowledgeable workforce. It is imperative that SLAD document its tribal knowledge before its senior personnel retire. SLAD would greatly benefit from a strong professional development program that is rich in training opportunities and participation in workshops, conferences, and continuing education (e.g., modeling and simulation workshops). SLAD would also benefit from hosting conferences and workshops in its areas of expertise. Although it has demonstrated improved knowledge of the literature and more collaboration, SLAD is still not a leader in most of the relevant technical areas. Active leadership in the greater technical community would serve to enhance SLAD's reputation both within and outside the Army and its ability to recruit bright scientists and engineers.

Other possibilities for professional development exist within current SLAD programs. For example, the program on developing a metric to predict mild traumatic brain injury represents a new line of endeavor for SLAD; it may not culminate in a fielded metric, but it adds to the scientific knowledge base. The program is impressive in terms of its collaboration and participation in the broader research community.

SLAD traditionally has been successful in shifting emphasis as appropriate to address the changing demands and requirements of technology and DoD strategy. Information operations and cyberwarfare are examples of recent changes that significantly affect SLAD. SLAD has extended its portfolio to include assessment of vulnerability of communications, networks, and information processing on the battlefield. Current and future focus will have to include more modeling and simulation applications, integrated with physical testing and evaluation, to include more predictive capabilities and risk analyses in these areas.

SLAD continues to extend its suite of analytical tools by building on successful collaborations and by capitalizing on the success of customer-driven, short-turn-around-time assessment programs to identify technical gaps that would benefit from enhanced tool development. The overarching umbrella for tool development is SLAD's system-of-systems analysis (SoSA) tool, S4. Most of the tools being developed within SLAD can be used as stand-alone tools, but most are also slated for use within the model library of the S4 code. The many challenges that surround this plan of action are discussed separately at the end of this chapter. This section discusses the challenges and opportunities associated with enhancing individual programs and tools.

In certain programs, such as the one examining lethality and vulnerability for the ground Combat Vehicle Program, SLAD estimates and assesses the impact of changes against known and projected threats. SLAD's record in meeting such challenges is good. The challenges include delivering results as quickly as possible. Any improvements in responsiveness would be appreciated by developers and decision makers.

The question of whether to buy or develop software is often faced by SLAD analysts. In assessing the vulnerability of various electronic weapons, the challenge is to come up with a code to convert from commercially supplied solid geometry models (supplied by contractors) directly into SLAD vulnerability models to achieve radio frequency fidelity commensurate with model requirements. SLAD should examine the commercially available software for providing this conversion.

MUVES 3 is the next-generation ballistic vulnerability/lethality model software code developed to replace the well-used MUVES-S-2 code. MUVES-S-2 will eventually be retired, which will require that all users transition to use of the newer MUVES 3 code. The plan for transitioning from MUVES-S-2 to MUVES 3 does not appear to demonstrate involvement of current MUVES-S-2 users outside of SLAD but still within the Army—specifically the Army Materiel Systems Analysis Activity (AMSAA) and the Center for Army Analysis and the Training and Doctrine Command Analysis Center (TRAC). These organizations depend on data produced by MUVES-S-2 and need to be considered in the transition to MUVES 3. This transition program should be expanded to support all organizations that depend on MUVES-S-2 output.

The program to develop a metric to predict mild traumatic brain injury is a new program for SLAD and much remains to be done to establish the value, both in terms of medical and operational feasibility of the proposed brain injury metric. Consideration of cognitive effects should be made a clear part of the research plan. As the range of relevant research grows (for example, automotive research and sports medicine), it will be a challenge for the currently sized research team to keep apace of developments.

Knowledge of the pertinent literature is imperative for the success of any solid research program. SLAD presenters are providing more discussion of the relevant literature, but awareness of the literature remains lacking in several program areas. SLAD is to be commended for moving into the important area of assessing the combined performance of materiel and humans. Army analyses typically consider the performance of a vehicle and that of its operators independently. Such separate treatment often misses critical issues arising from interactions between them, especially if either is in a degraded state or a stressing environment. A broad literature on workload analysis and related work stress could conceivably relate to driving in combat operations. This work dates back to the early 1950s and considers vigilance, eye-hand coordination, physiological measures, and psychological stress levels during various environmental conditions. An extensive review of the previous pertinent work is needed in order to ensure an appropriately scoped niche for SLAD. SLAD should expand interdisciplinary efforts with other organizations to include human performance groups at appropriate universities and laboratories. For example, the design and development of the RAH-66 Comanche were addressed in a manner that integrated attention to humans and equipment, which included understanding of the man-machine interface, reduced pilot workload with single-pilot flight capability, crew cockpit ballistic protection, and crew cockpit crash worthiness protection.

SLAD is conducting a series of UBB projects aimed at refining and validating high-order models that assess UBB vulnerability, in order to develop reduced-order models that are quick-running and focused. Both the approach and SLAD expertise are appropriate. Emphasis should first be placed on completion of the high-resolution model and its validation. Although there is merit in having a methodology for strategically allocating capabilities for homing in on specific areas of concern, there is also risk in losing the ability to detect what could be important interactions. Caution should be applied in the experimental design of analyses using the reduced-order models. The value of being able to participate early in the development cycle (e.g., in requirements determination or concept evaluation, when data packages are high-level and low-resolution) suggests that the directorate might consider a reduced-order version of many of its models to identify unattainable requirements and filter out infeasible or unresponsive alternative concepts. The appropriate degree-of-order reduction will, of course, depend on the application. SLAD recognizes the value of early participation and is developing appropriate methodology.

The good work of SLAD analysts is rarely evinced at early phases in the design process of equipment and components. SLAD presented many examples in which its analyses made substantial contributions late in the acquisition process. The Army would benefit greatly if these significant contributions in equipment evaluation were made much earlier in the acquisition cycle. Examples of these programs include the following:

- *Initial experiences in testing and assessing active protection systems for light-armored vehicles.* This project's goal is to determine performance of proposed systems. The project provides evidence of SLAD's core contributions to the acquisition process, but it would have been even more effective if SLAD had been engaged earlier in the life of the acquisition program. This very good work reflected what SLAD ought to be doing, and it resulted in rapid deployment, which positions SLAD for involvement in the upgrades and design cycles of Army systems.

- *Analyzing combat injuries to drive the development of body armor.* This effort is an excellent demonstration of the benefit that SLAD provides to the Army and specifically how SLAD can improve the Army acquisition process through earlier involvement.
- *Reducing vulnerability from damaged lithium-ion batteries.* Results of this effort should be embedded in requirements for any new vehicles proposing the use of similar batteries. SLAD should capture and share data with both industry and other government agencies doing work in this area to expand the knowledge base.

SLAD is developing valuable insights and analytic capability in fields that are critical to modernizing the Army. Making prospective customers, including those outside of the Army, aware of SLAD capabilities will be important to establishing a role for SLAD that is analogous to its preeminent work in kinetic survivability. Making those in the Army acquisition process aware of SLAD capabilities will help to ensure that SLAD is established as an influential partner in equipment design and acquistion.

SLAD has many state-of-the-art test facilities that support its test and analysis capabilities. Expanding community awareness of these facilities will aid in building collaborations inside the Army, with other DoD laboratories, and with academia and industry. The Electromagnetic Vulnerability Assessment Facility (EMVAF) is a state-of-the-art facility in which leading-edge experimentation is being conducted (e.g., GPS jamming). Few Army laboratories currently perform this testing on dismounts, so SLAD should pursue greater visibility and collaboration for this effort. During discussion of the electronic warfare (EW) methodology development project, SLAD identified an anechoic chamber and a reverb chamber. No clear delineation was made between the capabilities of these chambers and other competing chambers in facilities outside the Army. SLAD should benchmark all of its facilities to show how they compare with other facilities for end-to-end testing. SLAD should consider showcasing all of its unique and state-of-the-art test facilities; methods might include hosting booths at DoD and other professional conferences and publishing papers that describe the research that SLAD is conducting in these facilities.

A facility to bench test and satisfy DoD T&E requirements for a system as large as a helicopter is expensive, but the Rotorcraft Survivability Assessment Facility allows the completion of live-fire testing with more precision and less risk than any other facility. The challenge will be to increase utilization and show the overall cost savings.

Many SLAD programs offer the opportunity for collaboration. Although SLAD has shown great improvement in the number of collaborations, there is room for expansion. Examples of areas ripe for collaborative efforts include the following:

- *Computer network operations.* SLAD needs to secure participation at the war games and or training environments (e.g., the EUCOM war-game in Stuttgart).
- *INVA/DE.* The team should establish a relationship with NETWARCOM.
- *Optical augmentation (OA) and SLAD theoretical OA modeling.* This program would benefit greatly from collaboration with a software developer/coder. SLAD could gain information by examining what the Air Force program in this area is doing.
- *Lethality analysis for medium-caliber munitions in urban environments.* The project demonstrated awareness of previous U.S. work in urban combat but did not uncover available data from work done by our Allies, specifically German efforts in military operations on urban terrain. An opportunity for additional data on wall performance and definitions may be available within the field of seismic research. Urban environments are very complex and can alter expected weapon effects. Because this project addresses a deficiency perceived within the DoD, identification of other ongoing efforts and collaborations within the DoD should be identified.

- *Lethality and vulnerability analysis for the Ground Combat Vehicle Program.* SLAD should expand collaborations with universities and industrial laboratory research programs that support vehicle manufacturers and crash testing programs. SLAD should capture relevant lessons learned from Iraq, Afghanistan, and other fields of action. In addition, SLAD should understand what JIEDDO and other rapid response organizations are doing to address related issues. This is an opportunity to position SLAD to influence the requirements for new combat ground vehicles.

OVERALL TECHNICAL QUALITY OF THE WORK

SLAD aims to provide sound assessment and evaluation support of the survivability, lethality, and vulnerability (SLV) of Army equipment and soldier systems to ensure that soldiers and systems can survive and function reliably in full-spectrum battlefield environments. Although SLAD is not as well staffed as it should be, its staff is competent, creative, and enthusiastic about SLAD's mission. SLAD facilities include state-of-the-art laboratories and equipment that enable ground-breaking research and provide a strong basis for collaborations with outside partners. SLAD has many opportunities to capitalize on its test and evaluation base to selectively expand programs, to develop tools and methodologies to broaden its analysis capabilities, and to define and maintain the competitive edge essential to being the Army's primary source for SLV assessment.

SLAD does an excellent job of assessing and researching tightly constrained instances of vulnerability and survivability. With each step that relaxes those constraints by expanding the organizational, temporal, and spatial envelopes addressed, SLAD's ability to develop and apply appropriate methodology and analytic capability is increasingly taxed. It may now be overextended and thus less efficient and effective than necessary. SLAD should prioritize its development of modeling and analysis tools to ensure adequate allocation of resources to areas that it can most effectively support.

Physical testing is expensive and time consuming. Therefore, it makes sense for SLAD to construct high-resolution models of certain phenomena and, once they have been validated by physical testing, to use the models in place of subsequent physical tests. This work requires engineers, physicists, and analysts who are comfortable with the fundamental principles of the physics involved. In most cases SLAD demonstrates a solid understanding and sound application of scientific principles in developing engineering solutions to applied problems, but there is some evidence that not all SLAD personnel have a comfortable working knowledge of sound physics principles.

It is not necessary for SLAD to become involved primarily in basic scientific research to be a viable contributor to the Army and scientific communities. SLAD is uniquely positioned to observe emerging Army needs because of its core competence in rapid response programs. SLAD uses information produced by many of these small, customer-driven programs to identify areas ripe for research and tool development. Although gap analyses and niche identification are improving, the priority of these fundamental aspects of program planning should be elevated. Clear identification of the gaps in relevant research is the first step toward ensuring that SLAD programs are well positioned to contribute in areas that take advantage of SLAD strengths and provide the context in which SLAD collaboration would benefit advancement of the state of the art.

One of the largest challenges that SLAD faces as it moves toward greater emphasis on modeling tool development is hiring a sufficient cadre of appropriate personnel. With about 45 percent of its personnel over the age of 50 and only 8 percent of its scientific and engineering workforce possessing Ph.D.'s, SLAD needs to bring on board new personnel and train current employees in new areas. SLAD should consider looking toward postdoctoral, sabbatical, internship, and visiting scholar programs to provide healthy intellectual input and invigoration. In the short term, SLAD should consider expanding its postdoctoral program to attract personnel trained in needed areas and providing continuing education of SLAD personnel through broader collaborations in academia and other DoD institutions. One avenue for accomplishing these objectives might exist through the National Research Council postdoctoral program, of which ARL is a member. SLAD should encourage and incentivize more personnel to pursue advanced degrees (M.S., Ph.D., and post-degree sabbaticals) in relevant areas. Partnering with the Army Research Office to fund programs, including 6.1 research and multi-university research initiatives (MURIs), in areas of interest to SLAD would help in training and recruiting personnel. Establishing centers of excellence in universities could also help to provide relevant training and experience to personnel already working in SLAD.

To ensure recognition of SLAD as a leader in the research, test, and analysis community, its personnel should document their work in open literature publications. SLAD has begun to publish a few peer-reviewed publications, but publication should become a more standard practice at SLAD. Publication in scientific or technical peer-reviewed journals provides traditional independent evidence of the rigor, efficacy, currency, correctness, and technical quality of the work that would define SLAD as a leader in innovative test and analysis tool development methodology.

SYSTEM-OF-SYSTEMS SURVIVABILITY SIMULATION

During the past 8 to 10 years the the ARLTAB has identified many concerns regarding the development of System-of-Systems Survivability Simulation (S4) software code. The need for analyses of systems of systems is real, and elements of the tools and models that support such analyses are lacking—in particular, analyses of engineering and technology used to accomplish the generation, processing, and exchange of information that produces desired coherent behavior of units and formations. These analyses require varying levels of fidelity and detail; taken together, they need to represent the dynamics of combat between opposing forces, including movement, fires, consumables, command and control, and communications. Current combat models, for example COMBAT XXI or OneSAF, produce detailed traces of the combat process and dynamics on digital terrain. These traces include instantiations of tactics, techniques, and procedures as well as command and control and communications. In general they do not provide high-fidelity representations of physical processes, such as path losses or protocols used in network communication. Higher-fidelity representations are needed to properly carry out systems analyses focused on identifying and responding to deficiencies in connectivity or capacity, dynamic management of bandwidth, or the performance of algorithms that process and act on sensed data. However, that higher fidelity is not needed throughout the complete space for system-of-systems analysis; in fact, there is a tradeoff between fidelity and analytic efficiency.

SLAD can make contributions to system-of-systems analysis. Whereas past designs for survivability focused on improving armor recipes, designs now encompass passive and active armor and countermeasures, managed in real time by onboard information networks. In the near future, designs will include cooperative survivability among groups of vehicles Collection, processing, and exchange of information will be key processes, integrated to perform real-time decision making and employment of effectors. SLAD logically has responsibility for analyzing such systems of systems from the perspective of physics and engineering. In the case of ballistic vulnerability, SLAD was provided with the tactical context in which rounds were launched and impacted on a vehicle. So, too, in the case of emerging system-of-systems survivability suites SLAD should turn to agencies such as TRAC that focus on and excel at tactical and operational analysis and modeling for context and combat dynamics, including tactics, techniques, procedures, and operational employment. To be effective, SLAD should carefully bound its role and should focus on competently extending SLAD's technical, engineering, and scientific analyses within appropriate operational and organizational bounds.

Changes and Accomplishments

Changes and accomplishments noted during this review cycle include the following:

- SLAD improved the definition of scope for S4. The focus of S4 modeling applications has been reduced from brigade level to company level and below.
- SLAD recently instituted a formal software development process. This process includes:
  —Better articulation and definition of the SLAD vision for system-of-systems analysis,
  —Formulation of a verification and validation strategy and required documentation of results,
  —Participation of the Los Alamos National Laboratory in the evaluation of S4,
  —Completion of S4 proof-of-principle efforts for the Army Evaluation Center (AEC) and the Program Executive Office for Integration, and
  —Plans by AEC to accredit the S4 model.
- SLAD expanded collaboration efforts. Collaborations include Stanford Research Institute, Sandia National Laboratory, Army Research Office, Air Force Research Laboratory, Air Force Office of Scientific Research, Naval Postgraduate School, and Army organizations both within and external to ARL (i.e., Army Test and Evaluation Command, Army Evaluation Center, CISD).
- SLAD reported that the Army Test and Evaluation Command has concluded that SLAD's system-of-systems analyses provide insight beyond current modeling and simulation capabilities and address measures not answered in tests or other modeling and simulation applications.
- SLAD has trained software development staff in capability maturity model integration (CMMI) tenets and practices of process improvement. SLAD has also contracted qualified experts in CMMI to overlay its framework on SLAD's systems and to help to formulate a plan for improving the rigor of its processes as necessary. However, the SLAD scope is limited to self-assessment, without going through formal certifications.

Opportunities and Challenges

SLAD should consider a tactical pause to review and more carefully define S4's role and mission. This review should include consideration of collaborative efforts with DoD agencies responsible for tactical and operational issues as well as SLAD's role in providing high-fidelity analysis and tools focused on the physics of new and emerging survivability designs. Although SLAD has made some improvements in S4, the software and SLAD's application of it have exhibited flaws, and progress has not yet been significant enough to warrant endorsement of the changes as meaningful. As SLAD plans and implements a systematic software development effort, more substantive improvements are expected. In addition, SLAD should carefully reexamine its technical plan for collaborating with and supporting the efforts of other modeling activities within the DoD (such as TRAC), focus its system-of-systems analysis (SoSA) modeling on cases that demonstrate the usefulness of S4 to others, and align its resources with its plan.

SLAD should prepare a detailed flowchart to determine the structure and capabilities of the current instantiation of the model to enable a more structured, well-thought-out, long-term plan for where this program should go to support the SLAD mission and strategic plan. SLAD should modify its model and software development process to become compliant with software development standards employed elsewhere in the acquisition and analysis community, such as the Software Engineering Institute's capability maturity model integration and its associated certifications. To ensure that the team includes the necessary modeling skills and experience, SLAD should consider re-competing the contract for S4 model development. For the next review, SLAD should provide a description of a representative low-level function in ultimate scientific detail, including any parameters that should be estimated by subject-matter experts (SMEs) and the meaning of those parameters to the SMEs. A good candidate here would be the function of shooting at equipment or troops with artillery.

Process and Methodology

SLAD should clearly identify both near- and long-term goals and work with its customers to establish a well-defined set of success metrics against which to evaluate progress. If S4 is to become an effective element of SLAD, then needed is a concise plan that includes the rationale for the use of system-of-systems analysis and S4 in SLAD and how S4 will contribute to the SLAD mission. The modeling effort should include clear definition of the survivability questions to be addressed, verification and validation issues, and the degree of confidence in the output. The value added by the modeling effort should be through the integration of the information from the engineering models, testing, and evaluations within SLAD. If S4 is to become a major tool for SLAD, then the goal should be to gain a balanced competence and credibility that is comparable to the other functional groups within the directorate.

SLAD's plan should include milestones for the development process, implementation, and validation. Equipment, software, and certifications should be established, and collaborating partnerships should be identified. The S4 team should strive to become recognized as an independent analysis and evaluation organization that plays an important role in assessing and evaluating the survivability and vulnerability of military weapon systems. It is also necessary to build partnering and networking relationships with other modeling and simulation organizations, with SLAD's unique contribution being that of integrating information from other programs within SLAD and from other ARL organizations (e.g., the Sensors and Electron Devices Directorate) into the S4 modeling efforts.

SLAD requires a dynamic technical leader for SoSA and S4 who will assume clear ownership of all aspects of the SoSA and S4 program and will be responsible for guiding and directing its development in a direction that demonstrates benefit to SLAD analysts and the Army in general. SLAD should adopt a professional software and model development process for S4 including a development plan that clearly identifies the following:

- Key customers, critical requirements, and long-term and short-term goals;
- A specific approach for attaining the goals, to include
  —Program structure, framework, and description;
  —Detailed descriptions of individual models contained within the framework or to be added, and how this will be done; and
  —Inputs, uncertainties for all, and their treatment;
- A plan for software quality assurance and model verification and validation, to include performance metrics; and
- Identification of resource requirements and how the program is to be leveraged.

Personnel

Most of SLAD's core staff displays an impressive domain expertise. That expertise is not present in the S4 developers, who lack experience in system-of-systems analysis, development of large-scale combat models, and software development. As detailed below, they are weak in the knowledge, skills, and attributes required to ensure success in this program. There is an adequate base of subject-matter expertise to design process models of sufficient detail in the areas of SLAD's core mission, including the network operations examples described during the review. However, that level of expertise is not present in the S4 model development team, as evidenced by the lack of (1) knowledge of the current state-of-the-art in combat models; (2) a focus on temporal, spatial, and organizational scales sufficient for SLAD's purpose; and (3) compliance with standard software development processes, including the development and documentation of requirements, the specification of an architecture and its description via flowcharts, the development of a data dictionary, and the production of users' and programmers' manuals.

SLAD does not appear to have personnel qualified to design and create software to support analysis of military operations, either maneuver unit warfare or wide area security. SLAD does not have personnel with the training, experience, and expertise to conduct system-of-systems analysis of military operations, to play a role in deciding what models, simulations, and tools are required. A competent staff of modeling and simulation personnel, active in the military operations research and systems analysis community, should be developed and expanded. SLAD should place a military operations researcher (OR) into a position of influence over the military mathematical modeling. Personnel development should focus on obtaining a strong operations research staff with special skills in modeling and simulation. To achieve that, it would be important to present SLAD as an exciting place to be for scientists and engineers, offering good work with a clear and important mission. It would also be important to make SLAD known and respected outside its home fields at Aberdeen and White Sands. Exhibiting SLAD's work at conferences and other professional venues would also help to entice recent graduates to consider an ARL career.

To leverage software and modeling support, SLAD should build strong relationships with ARL and DoD modeling and simulation groups. SLAD should also expand the source and breadth of SMEs and access to more soldiers with operational experience. To establish a broader institutional system-of-systems awareness, SLAD should encourage and enable its personnel to attend, as observers, advanced warfighting experiments, simulation exercises, and command post exercises, the last of these at all echelons. In this way a foundation of experience could be built up over time and SLAD could become prepared to play its key role in collaborative system-of-systems analysis.

The current approach to S4 development is flawed. All of the S4 coders and modelers reside at the New Mexico State University Physical Sciences Laboratory (PSL); only analysts reside at SLAD. The current methodology for performing S4 operations involves SLAD analysts working with PSL to define a problem. PSL personnel then run the code and provide data to the analysts. To date, PSL has not documented its code. Furthermore, there is no program plan in place, no users' manual, and no descriptive report with all of the details. Because of this approach, SLAD does not *own* the code, which is imperative for S4 success. If SLAD is unable to hire coders to assume ownership of the development and maintenance of the code, then the directorate should demand documentation and drive the direction of development through construction and enforcement of a detailed program plan.

Scope and Collaboration

To ensure that S4 will become a useful tool for the system-of-systems analyses that SLAD believes to be a requirement of SLAD's mission space, it is important that the scope of the model be carefully chosen. The scope of engagements should be at the platoon and squad level, and certainly no larger than the company level, with interactions and dependencies on non-organic elements represented via exogenous events. It might be efficient for SLAD to consider S4 as composed of two components. The first, for which SLAD would be responsible, would be high-fidelity representations focusing on physical survivability and lethality as well as the sensing and information networking on which those processes depend. The second component would be the use of dynamic traces of combat, generated by models such as those used by TRAC, to drive the SLAD component. It is worth noting that this approach would permit SLAD to focus on appropriate slices in time and space and would thereby produce efficiencies in analysis.

SLAD should collaborate with those developing and using COMBAT 21 and other models (TRAC, AMSAA, Center for Army Analysis, and others) to gain relevant operational knowledge and increased military credibility. SLAD staff need to understand what is done well, how it is done, where deficiencies exist, and what is operationally important. Failure to include these other important stakeholders will result in analysis with no buy-in from customers of the operational community.

One promising way ahead for the Army would be for SLAD to work with TRAC-WSMR to create a joint model for mutual use. SLAD would then be starting from a supported object-oriented model that would presumably have established military credibility and could provide the basis for achieving higher fidelity. This would reduce SLAD's problems associated with adding modules with the increased fidelity that it believes necessary for its customers. This collaboration would allow TRAC to provide valid tactical information to SLAD and would allow SLAD to provide valid technical information to TRAC.

Other collaborators who should be considered include communities of practice and stakeholder groups (military, academia), of which the following are examples:

- The system-of-systems engineering community (for example, the Undersecretary of Defense for Acquisition, Technology, and Logistics has published the *Systems Engineering Guide for Systems of Systems*[3]).
- The modeling and simulation community and standards groups—for example, the Computer Generated Forces (conference), Simulation Interoperability Standards Organization, High Level Architecture, and the Synthetic Environment Data Representation standard group.

Specific Recommendations

- SLAD should consider a tactical pause to review and more carefully define the role and mission of S4.
- SLAD should plan and implement a systematic software development effort.
- SLAD should carefully reexamine its technical plan for collaborating with and supporting the efforts of other modeling activities within the DoD (such as TRAC), focus its system-of-systems analysis modeling on cases that demonstrate the usefulness of S4 to others, and align its resources with its plan.
- The S4 team should investigate integration of the ACQUIRE model to augment the S4 sensor models.
- The S4 team should develop a consolidated set of new communication features for S4 with input from the Army communications community.
- SLAD should have solid engagement with the intelligence community.
- SLAD should leverage the services of AMSAA for validation, verification, and assessment.

[3]Systems Engineering Guide for Systems of Systems, Version 1.0, August 2008, Director, Systems and Software Engineering; Deputy Under Secretary of Defense (Acquisition and Technology); Office of the Under Secretary of Defense (Acquisition, Technology and Logistics). Available at http://www.acq.osd.mil/se/docs/SE-Guide-for-SoS.pdf (accessed October 3, 2012).

- SLAD should focus on physics processes at a high-fidelity level (e.g., packet-level resolution in communications).
- S4 should be structured to accept combat dynamics as an input from other Army combat models.

Other concerns included the approach to survivability analysis with respect to information operations, the understanding and application of agent-based modeling, the need for strong information assurance and uncertainty analysis, and the lack of connectivity with relevant technical communities.

Given the importance of the soldier and experience in theater over the past decade, concentration on the soldier may be the critical research area in the short term. It is clearly an area where SLAD already has expertise.

Technical Merit

Progress has begun toward establishing a formal software development methodology and in bringing the system-of-systems analysis program, including S4, more under SLAD control, but there is still a long way to go to establish SLAD as a credible participant in the system-of-systems analysis environment. Relative to the metrics of quality assessment, the program does not reflect a broad understanding of the underlying science; the qualifications of team members are inadequate for the task; the facilities are state-of-the-art; the analytical work does not reflect a sound understanding of Army requirements; and the mix of theory and computation is appropriate, but the theoretical basis is inadequate.

6

Vehicle Technology Directorate and Autonomous Systems Enterprise

INTRODUCTION

The Panel on Air and Ground Vehicle Technology (AGVT), along with selected members of the Panels on Survivability and Lethality Analysis, Sensors and Electron Devices, Armor and Armaments, Digitization and Communication Sciences, and Soldier Systems, reviewed the Army Research Laboratory's (ARL's) Autonomous Systems Enterprise at the ARL facilities at Aberdeen, Maryland, on July 11-13, 2011. The Robotics Collaborative Technology Alliance (CTA) and the Micro Autonomous Systems and Technology (MAST) CTA are integral parts of ARL's Autonomous Systems Enterprise. The Vehicle Technology Directorate (VTD) was reviewed by the Panel on Air and Ground Vehicle Technology at its meeting on June 25-27, 2012. The Directorate has four divisions. Three divisions are aligned with the key scientific disciplines of mobility (propulsion, autonomous systems, and mechanics). The purpose of the fourth division—the Vehicle Applied Research Division (VARD)—is to provide notional concepts for the entire spectrum of Army vehicles. These notional concepts guide the Directorate's investment decisions by ensuring that all technologies required for a class of vehicles are covered and by defining key technologies that enable several types of vehicles. The June 25-27, 2012, meeting was the first review of VTD's new building and research laboratories called for by the 2005 Base Realignment and Closure Commission (BRAC). The review consisted of overviews given by management, presentations on a subset of current projects, and poster sessions at which project leaders were available for discussion.

VEHICLE TECHNOLOGY DIRECTORATE

Changes Since the Previous Review

Massive changes in VTD's manpower, research portfolio, and location have continued to occur since the 2009-2010 ARLTAB report was published. Perhaps most significant are a complete change in VTD leadership and completion of the new VTD building at Aberdeen.

The leadership change, coupled with the fact that a high percentage of VTD researchers have been hired only within the past 4 years, has caused a disruption in many research activities. In response to an ARLTAB recommendation to maintain a systems focus as it instituted changes in location, personnel, and research portfolio, VTD established the VARD. The VARD defined eight capability concepts vehicles that address Army objectives and critical capability needs (see Table 6.1). VTD began to align its research portfolio to address the technological requirements of the eight capability concept vehicles. However, the VARD appears to have lost focus on the eight capability concepts, and the capability concepts are not being utilized to guide the research portfolio. Good research is being conducted, but how this research fits into the VTD mission is not well understood. Therefore, VTD management should refocus attention on capability concepts as a systems methodology to align its research portfolio to meet critical Army needs and requirements.

Construction at Aberdeen Proving Ground of a 35,500 square foot building to house VTD personnel and corresponding research laboratories as required by the 2005 BRAC has been largely completed. Relocation of VTD personnel to the new VTD building is complete, and the development-checkout of

TABLE 6.1 VTD's Mobility Capabilities Concepts Approach

Capability Concept	Army Objective	Critical Capability Needs
Persistent Staring Intelligence, Surveillance, and Reconnaissance	Improved and persistent situational awareness for military operations	High-endurance VTOL Autonomous operation
Cargo Unmanned Aerial Systems	Overcome sustainment shortfalls associated with current supply methods	High-speed VTOL Autonomous operation Automated cargo handling
Multirole/ISR Attack Vertical Takeoff and Landing (VTOL)	6K/95 armed aerial escort with higher speed/longer range than current fleet	Variable-speed power/drive Adaptable rotor performance
Long-Range Heavy Lift	Mounted vertical maneuver into austere environments	Large stable rotor Large efficient propulsion Lightweight durable structure
Advanced Ground Combat Vehicle with Unmanned Ground Vehicle Wingman	Improved survivability and mobility for armored vehicles	Reliable efficient propulsion Armored robotic vehicle
Terrain Adaptable Tactical Wheeled Vehicle	Tactical transport with robust mobility in austere terrain	Reconfigurable suspension Advanced high power diesel
Small Dexterous Robots	Soldier tasks performed at a safe stand-off distance	Higher levels of autonomy Dexterous manipulation
Micro Autonomous Systems	Tactical situational awareness	Low-power mobility Distributed autonomous operations

SOURCE: National Research Council. 2011. *2009-2010 Assessment of the Army Research Laboratory.* Washington, D.C.: The National Academies Press. Page 71.

the new research laboratories is ongoing. A memorandum of understanding between VTD and NASA remains in place for the 11 VTD staff that will remain at the NASA Glenn or NASA Langley facilities.

VTD leadership has continued to transition the research portfolio from a focus on large helicopters to a focus on the entire spectrum of Army ground and air vehicles. This transition has required the hiring of many new researchers and the establishment of many new research areas. There is evidence that a lack of good senior mentoring of these new researchers has limited the usefulness of their research.

The two CTAs that have had continuity of leadership are in good shape relative to the rest of VTD research portfolio.

Accomplishments and Advancements

VTD's most significant accomplishments over the past 2 years can be found in the areas of the Robotics and MAST CTAs and the start-up and checkout of the new VTD laboratory areas.

Small Engine Altitude Performance and Heat Engine Systems Altitude Test Facility

This state-of-the-art facility is unique relative to those of other government laboratories. Overall, the facility is well designed and provides diagnostic capability for testing a variety of engine types and environmental conditions. The facility has the capability to simulate sea level to 25,000 feet, provide air from –40 to +130 degrees Fahrenheit, and has two dynamometers, which together cover a span from 1 to 250 horsepower (hp). The current work is focused on small engines of 40 hp or less under a subatmospheric variable-pressure environment, where engine performance and components (e.g., bearings and seals) will be evaluated. The testing of a Wankel engine is planned for the future. Combined with computational analysis of engine performance with a gas turbine suite of commercial modeling software, this testing will provide VTD with the ability to test all types of Army engines. Planned testing of a Shadow unmanned aerial vehicle (UAV) engine in a French facility will provide stand-to-stand validation of this facility. ARL should support the purchase of the needed ancillary equipment to support the stand. The goals of the experimental effort should be clearly defined to fit the VTD goals of developing advanced modeling and a complete fleet of engines that can operate on JP-8.

Gear Box Transmission Test Facility

Although the test equipment for this facility is not yet in place, its description suggests that this is a unique, much needed Army facility. A VTD presentation discussing the Army's experience with poor efficiency and reliability of hydroid drives provides an example of why this facility is needed.

High-Temperature Materials Development Facility

This facility contains the standard burner rig type of equipment that can be found in many laboratories across the United States. VTD has need for high-temperature materials in many if not all of its capability concepts. Although it should be involved in high-temperature materials research given the serious flaws in such work conducted by the NASA Glenn Center for VTD, VTD has determined that other work is of higher priority and therefore has appropriately postponed developing a high-temperature materials program.

Combustion Facility

The Army's objective of utilizing JP-8 fuel in all of its vehicles makes this facility a necessity for VTD. The utilization of JP-8 over the range of engines of interest to the Army is a formidable challenge, the depth of which VTD may not fully understand. JP-8 combines the worst characteristics of diesel and gasoline. The high combustion chamber surface area relative to combustion volume makes it especially difficult to run Wankel engines and small diesel engines on JP-8. The utilization of small injectors and heated fuel has been unsuccessfully tried by others in the past. Perhaps VTD should also add spark assist and catalytic breakdown of the JP-8 to its portfolio to investigate this difficult problem. The current combustion facility allows for some interesting experiments; however, it should be modified to cover the entire range of sizes and combustion configurations of interest to the Army (e.g., Wankel combustion). In addition, the facility should be modified to measure temperature and velocity in the combustion process. Also required is analysis of what sprays are needed in what engines to obtain the required mixture-ratio distribution by the penetration of the air by the droplets. A clearer explanation is also needed about the relation of the current experimental facilities to these various configurations. In a similar manner, analysis should be performed to indicate where vaporization is sufficiently slow to control the overall combustion rate. VTD should consider how all of the concepts discussed, such as flash heating of the fuel, fit into an overall program aimed at delivering JP-8 combustion over a range of engine varieties and sizes.

Tribology Facility

The tribology facility contains many of the test rigs that can be found in laboratories across the United States; however, this facility clearly fits a VTD need. VTD should develop a fully defined program to address the Army's needs in this facility.

Low Speed Wind Tunnel

Because many of the Army's needs involve air and ground vehicles at relatively low speeds, this facility fills a VTD requirement. However, plans to test the hover flight conditions and slow forward flight of micro air vehicles in this wind tunnel are not likely to yield realistic results. The size of the test section will cause the downwash from the lift jets to be reticulated by the proximity of the tunnel walls. The utilization of a free jet may be required to accurately test the micro air vehicles. ARL should support the purchase of velocimeter equipment necessary to fully check out and test vehicles in the wind tunnel. The turbulence level in the test section of the tunnel has not been measured. There is risk that the turbulence levels stated for the tunnel will not be obtained in the tunnel. The addition of upstream screens and honeycomb has been utilized by many people to lower wind tunnel test section turbulence. VTD management should obtain help from outside experts before undertaking such an effort.

Continuous Trailing Edge Flap for Helicopter Rotor Blades Research

This is a well-conceived and well-designed program using a novel approach—embedding a piezoelectric actuated biomorph in a rotor blade for the purpose of trailing edge deflection. This concept offers an alternative approach to discrete flaps, which are commonly researched for application to helicopter rotor blades. An advantage claimed is that the proposed flap structure is simpler than a discrete flap because there are no mechanical components and moving parts. Additionally, a continuous blade flap structure is considered to be more efficient because there is no gap between the flap and the primary structure as found with discrete flaps. The system offers the opportunity to eliminate the swashplate as the means of primary flight control, thereby reducing helicopter weight, drag, and maintenance requirements. The concept is being developed using structural and aeroelastic analyses for design and bench and wind tunnel tests for model validation. Researchers are challenged with designing a trailing edge with sufficient flexibility to obtain large trailing edge deformation, but with sufficient stiffness to retain structural integrity, that is, to prevent any undesirable aeroelastic instabilities.

Ducted Rotor System Research

This work consisted of using computational fluid dynamics to predict the effect of diffusing the slipstream of a ducted fan. The researcher observed that the thrust decreased as the flow was diffused, rather than increased as expected, and the experiment was considered a failure. In fact, this result should have been expected, because the static thrust of ducted fans is increased by converging rather than diffusing the flow. This is another case when senior mentors would have been of great help to the researcher.

Opportunities and Challenges

The most critical challenge facing VTD is a complete lack of a plan for its research portfolio to meet critical Army needs. Without this overall plan the new research facilities do not have focus, and the research by the talented new researchers will not produce the critical technologies needed by the Army. Elements of this lack of a VTD plan have been evident for several years; however, the move to the new building with the new research facilities, the complete change in management, and the large number of new researchers make the lack of a plan critical at this time. Capability concepts are a tool to help with this effort, and it is extremely disappointing that VTD seems not to be using these concepts. The elements of the plan should start with critical Army needs and timelines. These needs should then be mapped into programs for each of the new research facilities, and a complete review of the VTD portfolio of research should be performed to determine what areas need added emphasis, what research areas are missing, and what research areas need to be stopped. The quality of VTD research is in decline and will continue to decline until an overall plan with timelines to meet Army critical needs is put in place. It is imperative that VTD's management put in place within the next year an overall plan with timelines.

New researchers are working with limited mentorship from new management and senior researchers. An example of this lack of senior mentorship can be found in the work on physics of failure from multi-dynamic axial excitation. This research dealt with the physics of multi-axis loading and explored the advantages to doing vibration/dynamic load testing in a uni-axial superposition method versus multi-axial loading. The research was both good and valuable, although the results were not well understood by the researcher. The research showed vastly decreased time to failure in multi-axial vibration testing. However, it also confirmed that uni-axial results could be accurately superimposed on each other to get a similar result. The researchers did not seem to grasp that the results showed that if one is not going to go through the modeling and stress superposition exercise for uni-axial testing, then the only way to get an accurate time to failure is via a multi-axial testing technique. Essentially what was demonstrated was that if one would take the time to model the superposition of stress generated by uni-axial excitations, there would be little advantage to doing the testing in a multi-axial manner. All of these are good and useful results; however, the researcher's conclusion that the uni-axial superposition method would not work is incorrect. This example, along with several others, spurs the suggestion that VTD management should utilize workshops and consultants to obtain the help of outside specialists who can mentor its array of new researchers.

The establishment of the VARD was the first important step to ensure that VTD focus on critical Army needs. In order to meet the needs of the Army, VTD needs to ensure that all the technologies required to support each of the Army's spectrum of vehicles are contained in VTD's portfolio of research or that VTD knows where it being conducted elsewhere in the United States. A review of the needed technologies would help VTD management to identify highly important technologies that support more than one capability concept and technologies in their portfolio that do not support any capability concept. Then the challenge would be to ensure that each researcher can relate how his/her research supports one or more of the capability concepts. VTD management should have VARD develop or adopt a complete set of capability concepts to meet the Army's spectrum of critical needs and should conduct a review of its research portfolio vis-a-vis the needs of these capability concepts.

Laboratory support equipment is required to support a state-of-the-art vehicle laboratory. The new VTD research facilities have the potential to compose a state-of-the-art vehicle technology laboratory. However, a large group of support equipment is needed to meet the requirements of each of the major facilities. Examples of this type of laboratory support equipment are a laser velocimeter, hardness testers, high- and low-cycle fatigue testing equipment, microscopes, and boroscope equipment. ARL management should ensure that the required support equipment is purchased.

Several examples demonstrate that VTD lacks ownership of relevant Army problems. For example, a researcher working on the wear of hydroid drives focused only on the piece of the problem for which he has expertise. However, the real problem may be the surface finish of the gears or the windage of grease in the gear sets. Before launching a research effort, VTD management needs to ensure that the most important aspects of the problem are being addressed. Once again, senior mentoring and leadership would help to address this problem, but making sure that researchers own an entire higher-level problem, not just a piece of the problem, would also help to eliminate this lack of ownership.

Overall Technical Quality of the Work

Basic VTD Research Quality

The BRAC decision to consolidate VTD at Aberdeen, coupled with VTD management's focus on Army needs, has increased the quality of the VTD research portfolio. In 2010 the establishment of eight capability concepts that embody the technical breakthroughs needed to meet critical future Army needs was a major step toward focusing and upgrading VTD research. However, failure during this review cycle to continue to align the VTD research portfolio with these capability concepts represents a major lost opportunity. Focus can be lost when an entirely new management team has the pressure of establishing and moving into a new facility; however, it is critical that VTD management quickly refocus on the technologies needed by the capability concepts. For example, VTD recognized in 2010 that a crawling bug type vehicle should be added to the micro autonomous vehicle portfolio of research, but 2 years later it has not been added. Similarly, combustion of JP-8 in the small volume of small engines is an example of a crosscutting technology area that would impact several capability concepts and is therefore a high-priority research area. While a new combustion facility is part of the new VTD, building modifications to the facility will be required to support critical Army needs in this very important area. This also represents another area of lost opportunity. Without an overall VTD plan to meet critical Army needs, the overall quality of VTD's research will continue to degrade.

Quality of Individual Research Projects

The project on a continuous trailing edge flap for helicopter rotors evinced the highest quality among those presented and is of the caliber of the best in its field. Aspects of other projects, such as physics of failure from multi-dynamic excitation, separated flowing using nano-second pulsed plasma, and ducted rotor research, are important, but overall those projects are not of the highest caliber. Of inadequate quality are research projects into material damage precursors in composite structure, variable speed power turbine research, compressive sensing robust recovery of sparse mechanical signals from incomplete measurements, and slowed rotor unpowered take-off.

Autonomous Systems Enterprise

A detailed discussion of the Autonomous Systems Enterprise is presented later in this chapter. Many, but not all, of the research projects in autonomous systems are of the highest caliber; the combined quality of the research contained in the CTAs is cutting edge.

AUTONOMOUS SYSTEMS ENTERPRISE

This section first provides an overall assessment of the Autonomous Systems Enterprise and then highlights the four subject areas reviewed—microelectronics and perception, intelligence, human-robot interaction, and manipulation and mobility.

Overall Assessment

Accomplishments and Advancements

ARL should be complimented for its utilization of CTAs that have allowed ARL to connect with and leverage world-class research in the area of robotics. These CTA activities are unique in the history of government laboratories, and without them ARL would not have been able to pull outside expertise into ARL. Perhaps ARL's most important contribution has been the use of the Micro Autonomous Systems and Technology (MAST) and Robotics CTAs to maximally leverage high-caliber research and talent across the United States, which is a significant accomplishment. Another benefit is the sense of community and excitement about robotics of many bright early-career researchers involved in the programs.

Opportunities and Challenges

In terms of balance, the research portfolio covers the over-the-horizon area very well. More attention should be given to near-term accomplishments. In a similar manner, the portfolio is too heavily weighted toward analysis; more experimentation would provide a better balance.

ARL should utilize two systems-oriented strategies to connect near-term needs to far-term research: (1) a road map of specific objectives, tasks, and challenges to reach the far-term goals and (2) off-ramps from this road map to achieve mid-term results. One example of these off-ramps is the TARDEC Appliqué project, in which the Army's ground fleet was converted from being manned to being optionally manned.

Enterprise management did not present an integrated long-term plan (road map) showing how the individual research projects that make up the ARL Autonomous Systems Enterprise fit together, or how and when ARL plans to transition this research into demonstration and/or development projects. Consequently, some of the research tasks have little or no connection to other research tasks taking place under the rubric of the enterprise. ARL should develop an integrated long-term plan (road map) that places additional emphasis on technology demonstrations.

ARL's role in developing the necessary 6.1- and 6.2-type technology for the future is extremely important to the success of Army systems. Alignment of the ARL robotics enterprise and overall Army strategic plans is not clear. Also not clear are whether the various ARL technology developments can be categorized as unique, essential, or relevant to the Army and how those developments position ARL and the Army for the near, mid, and long term. The following questions need to be answered: What gaps exist between the civilian and military markets? What can the Army leverage? What is missing? What will ARL provide? What will collaborators provide? Improvised explosive devices (IEDs) were stated as one of the main foci of the enterprise, but no significant discussion of IEDs was presented.

The ARL process faces a difficult challenge in grouping technologies according to near-, mid-, and long-term needs. A second and perhaps more difficult challenge exists in ensuring that adequate funding and a management structure are present to achieve cross-technology, integrated solutions. A third difficult problem relates to the application of revolutionary breakthrough technology solutions, arising from inventiveness of scientists and engineers, which cannot be planned.

Reducing the tendency toward stovepipe solutions is a frequent challenge for scientific and technical organizations. ARL should balance its efforts across disciplines, ensure that resources are moved when technology transitions, and maintain strong participation in the overall effort. Long-term requirements should be addressed primarily by 6.1 efforts, which should reflect the need for both capability-linked programs and novel concepts or breathtaking technologies. These programs do not lend themselves to rigid milestones, and allocation of resources has to be based on opportunity, scientific judgment, and a vision of potential applicability to Army systems. ARL should determine what sharing is occurring, among whom, and what joint investments are being made.

Commercial off-the-shelf robotic systems represent quantitative benchmarks against which the ARL should always measure its programs. Because some of the currently available systems are competitors with respect to what ARL is attempting to develop, at a minimum the option to develop or buy should be evaluated. Needed are metrics of success or failure for projects transitioned, stopped/divested, and ongoing.

ARL did not present research efforts related to the deployment of remote weapon systems on unmanned air or ground platforms; yet the Army has a need for armed robots that can effectively engage threat combatants and other targets to help improve soldier survivability. The expected future Army employment of armed robots suggests a number of new research questions that ARL should consider addressing within this enterprise. Examples of relevant research topics include the following:

- What types of robotic weapons are most suitable for deployment from each class of unmanned air and ground platform;
- What issues are associated with firing a robotic weapon system on the move (and also from a stationary posture) and with being able to effectively engage a stationary or moving/fleeting target;
- Whether robotic weapon systems should have an autotracker, and if so, what capabilities the autotrackers should have;
- What is the maximum soldier-to-weapon and weapon-to-soldier communication latency that can be tolerated under various engagement conditions, and how small latencies can be achieved with wireless networked communications; and

- What capabilities and features will future autonomous robotic weapons need and what new research will help to realize these capabilities.

Three materials-related research tasks were described in association with the description of the project on next-generation actuators and materials. These three tasks were apparently the only ARL materials research related to unmanned systems. The Army has a systemic problem with materiel weight, and unmanned systems are expected to be both weight- and size-challenged, because they will need to be transported to and from each theater of operation and some of these unmanned systems will also need to be carried into battle by soldiers. ARL has an opportunity to plan and undertake new research tasks focused on weight- and size-reduction of unmanned systems. Such research could potentially investigate new types of lightweight materials, structures, and components for future unmanned systems.

The following activities were not identified in the presentation of the set of ARL's autonomous systems enterprise activities and therefore appear to be missing from those activities; these should be considered:

- *Counter-robot activity*. If the enemy is facing robots, then what actions will it take to counter the robotic threat?—for example, bat nets placed across windows and doors to trap flying robots. ARL should consider research approaches relevant to counter-robot activity.
- *IED sensors and activity*. Although almost all researchers spoke of IEDs, the research portfolio contains no research on IEDs. A large task force, the Joint IED Defeat Organization, is addressing the problem associated with IEDs, but ARL also should have some of its portfolio aimed at these problems.
- *Energy*. Having soldiers carry 10 lb batteries for 15 minutes of robotic operation is not good; combustion of JP-8 in very small engines allows a factor of 100 more power density than systems powered by electric batteries. Ways to achieve combustion of JP-8 in very small engines should be part of ARL's portfolio.
- *Transfer of power from source to ground*. The efficiency of robots in transferring energy of the power source to the ground is still two orders of magnitude less than that of biological systems. Work in this area should be included in the portfolio.

In terms of impact on Army programs, ARL should consider the need to capture lessons learned with respect to IEDs, understand the ideal distance of soldier-robot separation, and identify what can be learned about UAVs that is applicable to unmanned ground vehicles (UGVs).

Overall Technical Quality of the Work

Many, but not all, of the research projects in autonomous systems are of the highest caliber; the combined quality of the research contained in the CTAs is cutting edge. ARL has organized a significant effort to leverage ongoing research across the United States. The overall technical quality of the work is very high for each of the key areas addressed: microelectronics, sensing, signal processing, and perception; intelligence; human-robot interaction (HRI); and manipulation and mobility. Specifically, the scientific quality of the research is comparable to that executed at federal, university, and/or national laboratories both nationally and internationally. The overall research reflects a broad understanding of the underlying science and research conducted abroad. The research team's qualifications are very good. The facilities and laboratory equipment are state-of-the-art, and appropriate laboratory equipment and models are used.

The CTA exhibited a good mix of Army, university, and industry involvement. However, coordinating the research at different locations and enabling collaboration among members of the alliance can be challenging. The following can improve the quality of the research:

- A systems approach is needed. This is ARL's responsibility, not that of the academic researchers or industry partners.
- An element of the systems approach is termed "capability concepts." For example, the description of the MAST goals almost completely reflects a capability concept. That is, the MAST robot will weigh 200 grams, be utilized in the last 100 meters, and work for three different scenarios. To complete the MAST capability concept, time to complete the mission, reliability (in terms of getting there and back without breakdown and in terms of finding men and bombs when there), and availability of the robot to go on the mission should be added to the concept definition.
- Reliability and availability goals and demonstration for robots are keys to the issue of soldier trust in the robot; that is, to design robots that are trusted by humans, reliability and availability goals should be added to the research.
- Another example of the systems approach is the development of technology road maps tied to military need, identifying what is available now and what is needed, by when, in order to fill operational gaps. Greater emphasis should be placed on developing an overall robotics road map that defines the long-term Army needs and objectives and, therefore, individual program requirements. The road map should specify who will conduct the research and where it will be conducted.
- Another systems tool is the use of benchmarks. Benchmarking was illustrated with the demonstration of the PackBot robot, which weighs 47 lbs and uses 10 lbs of batteries in 15 minutes of operation. In this example 60 people ran the robot through one of the MAST scenarios. The MAST goals are factors of 10 or more greater than the benchmark, so how the PackBot capability will improve with time is an important question to consider.
- The MAST joint experiments are examples of system testing. However, they should be concentrated on demonstration tied to the MAST capability concept goals. The results of the experimentation should be crafted into a clear statement of where the system is relative to its capability concept goals. The Robotics CTA should adapt the joint experiment system for its program.
- The plan for how robots will go to war should guide research. For example, as robots get smaller they lose capability, and therefore small robots will most likely be used in swarms. At the start of WW II, the German strategy of how to use tanks was more important than the capability of the tanks themselves.
- Many of the individual efforts are aimed at achieving the same objective but are poorly coordinated. For example, the robotic mapping or world view is a difficult problem that warrants more than one approach. However, each researcher should be aware of and attempt to leverage the efforts of other researchers in the same space. In addition, to conserve resources, ARL should identify and select the most successful approaches.

Microelectronics and Perception

Overall, the ARL research portfolio does a good job of covering the size range of robots. Fully autonomous robotic operation may take 30 years to fully develop. Clearly as robots get smaller their ability to carry sensors and support large computing activity decreases; therefore, demonstrating fully autonomous operation on large robots and then moving to smaller ones is recommended. The MAST program is doing best-in-class work at reducing sensor size and weight. Off-ramps to divert developed technology into existing platforms should receive more emphasis. Also, scalability over robot platforms seems to be a major concern. There is a need to address scalability challenges with regard to capabilities (e.g., fidelity of the sensors, range and detection).

The selection of sensors to accomplish the perception element of an autonomous system should be driven by the task and mission to be accomplished and the objects, activities, and events that one is trying to find and characterize. Bounding the problem in this way leads to identification of what can be observed and a selection of the sensor suite and measurements associated with the set of observables. There was little discussion of the observables needed and the justification for the selection of sensors and associated processing.

Micro Autonomous Systems and Technology

The goals of the MAST program include reducing the sensor size, weight, and power by a factor of 100. The hair inertial sensor is projected to be 1.5 mm × 1.5 mm, which is approximately 4 times smaller than the conventional microelectromechanical system (MEMS) accelerometer.

The millimeter wave (MMW) radar work provides an approximate 100 times reduction in size, weight, and power with improved performance; the research resulting in this revolutionary technology development is of the highest caliber. An opportunity and challenge is to characterize and measure the effective range of this device for military utility. Once this is taken into account, the impact of the device can be accurately assessed. It is also important to illustrate more clearly how the work fits into an overall roadmap. The necessary next step in the maturation of the work is a technology demonstration as part of the future joint experiment to validate the utility of this technology. This work needs a comparison to the state of the art.

Signal Processing

The presentation on super-resolution processing described a technique for super-resolving three-dimensional (3D) range images acquired using a flash LADAR sensor for robotic applications in an urban environment. Work was done in collaboration with Carnegie Mellon University. Work was performed on the SwissRanger SR-3000 flash LADAR device, with a pixel count of 176 × 144 and 850 nm light modulated at 20 MHz. The system acquires range images at a maximum rate of 50 frames/second and has a field of view of 47.5 × 39.6 degrees. ARL's super-resolution algorithm was used on the commercial off-the-shelf device to illustrate the benefit of super-resolution for flash LADAR imagery. For this work, the algorithm was operational in post-processing mode; however, it can potentially be implemented to run in near-real time on a robotics platform. The illustrative example of the operation of the algorithms is overly simple.

The purpose of perception tasks is to reason about what is in the environment and what things in the environment are doing. Although the team has made progress in these areas, clarification of the extent of progress would be assisted by clearer articulation of the state of the art. Semantic understanding of static areas, particularly terrain and object classification, works very well over large data sets. Semantic understanding of dynamic areas (e.g., activity recognition) works reasonably well on small data sets and on a restricted set of activities. Work on distributed and collaborative perception (multiple robots, robots and people) is in progress.

The team should consider active use of multiple sensor input. This would involve joint simultaneous training and classification with multiple sensor streams, not just fusing best outputs from classification on individual sensor streams. The team should consider testing on data sets that the learning algorithm was not exposed to. Although it is common to train on one part of the data and test on other parts of the data, it is not so common to train on one data set and test on an entirely new data set, which would add value to the work. If possible, the team should consider acquiring data from the field—perhaps utilizing video, LADAR, and radar footage from deployments in Iraq and Afghanistan. Future effort should address processing in real life for real-time decisions. There is a need to explore work being done by other organizations, such as the Air Force Research Laboratory, the Defense Advanced Research Projects Agency (DARPA), and industrial laboratories.

Semantic Perception

The purpose of this effort is to reason about what is in the environment—for example, whether a group of pixels is a car or a window—and what things in the environment are doing—for example, moving in a threatening way. If successful, then the system should be able to perform these tasks automatically, albeit with offline human supervision or input.

The presentation clearly captured the technical barriers that make this work challenging. Complex interpretation leads to computationally intractable (optimization) problems because perception is noisy; fully supervised training is unrealistic; and integration of non-sensory sources of information (e.g., domain knowledge) is difficult (though clearly important).

The presentation described, at a high level, the technical approach as a focus on objects and activities relevant to robots and soldiers—not general objects and activities—using contextual cues (e.g., external information and domain-specific information). The work aims to develop new learning and optimization techniques to make perception problems tractable. At present, the problems are intractable even when the focus is restricted to objects and activities relevant to robots and soldiers, and when context is built in.

The presentation provided a clear evaluation strategy for reviewing the state of the art and setting and achieving specific quantitative goals for the program. The goals include standardizing metrics and data sets (the data sets will be made public—a laudable goal) and producing publications.

The concept of semantic perception connects machine visual perception that is largely driven by physical objects in the visual scene with knowledge (i.e., domain knowledge, mission knowledge, and cultural information). This connection represents an advance because it allows context to drive expectations and reduce the space of possibilities produced by bottom-up visual recognition of edges and vertices.

The team has made clear progress in all areas. Although sensing was not addressed in the presentation, the following topics were addressed in some detail:

- *Semantic understanding of static areas*: terrain and object classification, which works very well over large data sets;
- *Semantic understanding of dynamic areas*: activity recognition, which appears to work reasonably well on small data sets and a restricted set of activities; and
- *Distributed and collaborative perception*: multiple robots, and robots and people, which is a work in progress.

Intelligence

Overall Assessment

Robotic intelligence is defined as whether the robot knows its location, what path to choose, and what the surrounding threats and opportunities are. This key grouping of capabilities needs to be addressed by technologies that have to be developed for autonomous robotic operation. The presentation of this work showed slow but incremental progress in this area. Because this area is very challenging, ARL has appropriately been applying many approaches to finding solutions. However, many of these approaches are not fully coordinated with other approaches in the same area, and therefore researchers have not leveraged the lessons learned and emerging results of the other efforts. For example, it was mentioned that robots and warfighters do not actually collaborate, which makes the role of the warfighter in this project unclear. Nevertheless, some of the approaches are sufficiently developed for standardization—for example, mapping inside a building should now be standardized, and additional research should be aimed at object identification inside the building. ARL should lead a mapping effort to ensure that all areas are covered by the research and that maximum leverage is being obtained from the different approaches.

Advancements and Accomplishments The robotics enterprise is addressing some critical subproblems whose solutions are necessary for increasing robot intelligence (e.g., mapping, path planning, machine learning, robot control, and architectures for cognition). The team has made several incremental advances, such as modeling the multi-robot patrol problem in a new way and using machine learning in a variety of ways to improve robot intelligence. The quadrotor control is very impressive; however, it relies on some strong simplifying assumptions (e.g., having perfect localization from a fully instrumented laboratory) that would make it very difficult to apply to real Army missions. It is difficult to point to particular advances that could change the game in terms of robot intelligence for Army applications, but this research is still solid, incremental work.

Opportunities and Challenges ARL should better define robot intelligence as it relates to warfighter needs. For example, what type of decision-making capability is needed? Relatively little was communicated during the presentation about the types of intelligence and decision making that are required to achieve the vignettes that have been outlined. A better vision of what is needed for Army applications would help to focus the research in intelligence. In particular, ARL should identify the forms of robot intelligence that are uniquely required for military applications but are not addressed by the significant amount of civilian work being done on robot intelligence.

For most of the research in intelligence, it is difficult to measure the accomplishments and progress. It is often not clear how the research fits into the state of the art or how the problems being researched directly affect an Army mission objective. More work can be done to benchmark the research, including the definition of short-, mid-, and long-term benchmarks of Army relevance. More solid demonstrations that are compared to benchmarks would support the case that relevant progress is being made. Also, the autonomous robot challenge is an excellent opportunity to demonstrate how to empirically measure progress. For example, when successful autonomous operation range increases from a few hundred yards to a few miles, measurable progress has been made.

The researchers should better elucidate the fundamental limitations of the models being developed, as well as better address issues of uncertainty and robustness throughout all the research tasks. This comment has broad applications. For example, in the MAST CTA, several of the intelligence projects and integrated demonstration did not acknowledge the size, weight, and power constraints or the sensing and computational limitations inherent in the MAST robot vision. All research tasks in the MAST CTA should directly address these issues.

Possibly missing from this research portfolio, especially for the MAST CTA, are power-aware computing, intelligent power management, and algorithms for managing bandwidth. In addition, some areas of important teaming research were not mentioned, such as the overall organization of the teaming (e.g., centralized versus decentralized intelligence across teams, hierarchical teaming, or other organizational strategies) and multi-robot decision making at the mission level.

Many individual projects address the issue of robot mapping; however, these project teams are not talking with one another, which raises the following questions: What mission objectives are being addressed by these different mapping projects? Why not perform comparisons across mapping projects for a variety of Army applications? Why not study whether the key strengths of different mapping algorithms can be merged into a single approach? Why not study the tradeoffs for each type of mapping for various Army-relevant applications? For example, multiple projects are under way to autonomously map out the interior of buildings, and these projects appear to be making solid progress. Because most buildings worldwide are of multiple stories, it is time to conduct a down-select of interior mapping systems and focus the resulting system on some stair-climbing approach such as was demonstrated so effectively in the autonomous stairwell assent project. It is very important to consider all stair-climbing concepts, such as tracked, legged, and wheeled platforms, in the overall program of autonomous mapping of the interior of multiple-storied constructions. The importance of instantaneous interior building mapping to the military mission remains somewhat dubious. How important would it be to map the inside of the building if it could be instantaneously and reliably determined that there were no combatants or other dangers inside it?

Overall Technical Quality The robotics enterprise is grappling with difficult problems that have been around for a long time. The enterprise's approach is to pursue many different algorithms for these problems, and then at some point in the future decide which is most beneficial for a given Army application. Research advances are being made to the state of the art, mostly of an incremental nature. Not all of the research approaches are well-justified (e.g., the hybrid mapping research, which pursues techniques based on the available sensor, rather than best approach to solving the problem). A better justification of the research approaches being pursued is needed in many areas of the intelligence research—particularly in terms of military requirements. Robot research in the civilian market continues to dwarf ARL's efforts. ARL should strive to fill the military-relevant gaps left in the civilian research, not duplicate that research.

Multi-Robot Persistent Surveillance Planning as a Vehicle Routing Problem

This work addresses a long-existing problem using a new technique, that of multi-vehicle patrolling. The objective is to solve the problem exactly. Comparing the approach to a baseline of the traveling salesman solution is commendable. The ability to generate solutions that vary the visit times is a nice side-effect, because it reduces the external predictability of the system. This research should be presented more carefully, recognizing that the approach is actually not generating a globally optimal result and that heuristic approaches can have value over exact solutions, specifically for solving large-sized problems in this probably intractable domain. Using this technique as a foundation for addressing the general task allocation problem with spatio-temporal constraints is an idea worth pursuing. Feedback should be added to the approach in order to close the loop on the control.

There is a lack of knowledge regarding similar past and recent relevant activities conducted by NASA, DARPA, and other government organizations. Additionally, this work should better address the issues of noise and robustness. References to the traveling salesman problem were refreshing, but if the power, weight, and energy challenges for a single robot are significant, then it would seem that the task of choreographing a team of robots for which we lack remotely relevant battery or energy technologies is premature. These problems can become quite intractable with a moderate increase in problem dimensionality, and there is a need to explore super-computing platforms in this context, especially the parallel implementation of heuristic optimization algorithms to solve the larger scale problems. It is not clear whether a practical problem formulation is possible that would admit an exact solution as proposed by this preliminary work. Nevertheless, this has taken positive first steps to solving a difficult problem.

Machine Learning for Robotics

An overview presentation outlined the various ways in which machine learning is being used to improve robot performance. Using machine learning to tackle the challenging problems in robot intelligence is a good approach with many avenues for fruitful study. The advocated approach for control learning, which begins with a model for nominal control and then applies machine learning to optimize the control parameters, is an appropriate way of applying machine learning. Storing a database of experience to assist machine learning is a useful way to gather data for machine learning, although much theory needs to be developed to determine how to generate and identify the right data for a given problem. In the area of designing for learning ability, however, other than properly instrumenting the robot with sensors and monitors, the main ideas are not clear. The presentations lacked sufficiently concrete technical information on all of these learning topics. Because the civilian market is already doing a lot of work in this area, ARL should focus its efforts on the aspects of machine learning that are uniquely military in nature and not the focus of civilian efforts.

Developing Hybrid Maps to Promote Common Ground

This research attempts to develop topological maps in indoor environments that can be represented in human-understandable terms. To date, the main contribution has been the development of an online approach that works in cluttered environments. The approach involves analyzing point clouds from a Kinect sensor and decomposing the environment into regions and portals.

Generating human-understandable maps has potential benefit to Army applications. However, using the Kinect sensor just because it is available is not an adequate justification for the selected approach. This presentation could have been strengthened if a better justification for the selected approach had been given. Old techniques that look at the ceiling to determine the likely 2D footprint of the walls are most likely a better starting point. As with the other mapping research, this work would also likely benefit from increased conversations among the various performers who are researching localization and mapping.

The focus was on mapping Western structures and their contents—for example, opening doors, climbing stairs, identifying beds, tables, and chairs. In much of the third world, interior openings may be covered with a blanket or other hanging, ladders may be used to access other levels, and furniture may consist of cushions, mattresses, and the like. Even in modern buildings robots may be stymied by leaving the door open and hanging a blanket up in its place. Robots designed for Western architectures may also be stymied by hanging beads in windows and doors or by installing window screens. These problems should be addressed.

Joint Experimentation of Distributed Mapping

This presentation reported on experiments in multi-robot simultaneous localization and mapping (SLAM) in indoor environments, as part of the MAST CTA emphasis on cooperation. In this experiment, a team of ground robots used laser, camera, and communications links to map out an indoor area. Robots recruited help at intersection points. The novelty of this mapping approach lies in the use of graphical models for data association and the use of structures (e.g., doors, signs, and posters) as mapping landmarks. The use of high-level features also reduces the amount of required inter-robot communication. The resulting accuracy of the system was reported to be approximately 1 inch. This research is solid and demonstrates some important SLAM and multi-robot capabilities.

However, the research approach does not consider the constraints specific to the intended small-scale MAST robots—that is, size, weight, and power—or limited sensing and computational constraints. ARL reported that the graphical model approach reduces the typical computational requirements of mapping from O(n2) to O(n lg n), which is an important contribution. However, the research needs to do a better job in general of explicitly addressing the size, weight, power, sensing, and computational constraints. The software developed for this demonstration requires robots at the capability level of the Robotics CTA, rather than the MAST CTA.

Human-Robot Interaction

Leveraging Human Cognition

Most of the robot mobility and manipulation research leverages the mechanics of biological systems (e.g., insects, birds with flapping wings), which is not necessarily the case for robot cognition. However, human cognition is a hugely parallel system, and almost no ARL research addresses massively parallel processing. For example, in the area of perceptual expertise, detection of IEDs and similar threats can be quite complex, yet humans do these tasks, and some humans do them better than others. Why not examine human skill as we do the flapping wings of birds to reverse-engineer the human mechanisms? Another area of ARL robotics work, radar signatures, could also benefit from an understanding of the types of cues that human observers use to identify threat. On the positive side, a few presentations harnessed human mechanisms. In the human robotic interaction group of the Robotics CTA the areas of robotic trust and teamwork are benefiting from an understanding of human trust and teamwork. The work on neurocognition envisions the use of EEG signals to identify anomalies in the environment to which an expert detector would attend.

The work on neurocognitive application to robotics highlights a broad range of robotics applications based on the use of physiological measurements. These applications range from improving robot performance based on recorded human responses to varying a soldier's tasking based on assessment of fatigue stress or other mental state indicators. The robotics enterprise works closely with the neuroscience group at ARL's Human Research and Engineering Directorate (HRED) to focus the research and technology maturation in a direction that has a potential for transition to useful Army applications.

Human-Systems Integration and Function Allocation

Human-systems integration (HSI) is much more that HRI or human factors for robotics. HSI is systems engineering with humans at the center. Timing is very important; one cannot adequately engineer a system or integrate the human in the middle or at the end of system development. Human considerations need to occur up front. In the case of soldier systems, researchers need to ask very early in the conceptualization and design of equipment or systems whether the soldier will carry it, use it, like it, and need it, and whether the soldier's mission will improve as a result of its use. This observation should be repeated in all projects. In the beginning, the question of function or task allocation also becomes important: What tasks are best allocated to the robot and what tasks are best left to the human? These decisions require input from the warfighter. In contrast, the enterprise seems to be pushing technology that the soldier is not asking for and does not describe, with supporting rationale, the functions and tasks that will be allocated to the robotic system and to the human.

Although ARL investigators talk with warfighters, this interaction provides only rudimentary information, because the warfighter does not often have the breadth of technological and scientific knowledge to know what is possible or where weaknesses reside. The discipline of cognitive engineering possesses a body of tools and methodologies for cognitive task analysis that provide systematic approaches to eliciting this type of information (e.g., see Pew and Mavor[1]). It also is possible that near- and mid-term solutions may allocate more tasks to the human, strategically postponing allocation of the functions that are more challenging to automate. Some researchers reported that they talked with operators and warfighters as they developed the concept of the robot as co-combatant, but these discussions did not happen at the right level and were ad hoc and undocumented. If operators and warfighters are driving requirements, then interactions with them should be structured to the best extent possible. For example, showing a warfighter what ARL is doing with robotic learning differs from asking the warfighter what kinds of things a robot should assist with.

The MAST CTA and the Robotics CTA should collaborate more, especially in the area of HRI. When this issue was raised during the presentations, the response was, in some cases, that HRI is being handled by the Robotics CTA, which implies that the HRI considerations happen in parallel to MAST research. However, these considerations should be better integrated from the beginning, or the end result may not be used by or be useful for the human.

As is often the case in large organizations, there are stovepipes that should be better connected. Coordination and knowledge sharing should be improved within the CTAs and between the MAST CTA and the Robotics CTA. It may be beneficial to leverage the work or lessons learned from the previous Robotics CTA. Additionally, connections should be forged across CTAs and the ARL autonomous systems enterprise, between CTAs and external industry and academic units, as well as across DoD.

Common ground can be better achieved by providing a common data set to share within and outside of DoD. This common data set would provide ground truth (e.g., location of threats) and therefore could be used to compare alternative technologies.

Expanding HRI Research

ARL's employment of simulation in its research is good and should be extended and leveraged. HRI's role should be expanded beyond testing of swarming robots to cover all aspects and stages of robotic research. Systematic tools for doing HSI needs analysis should be used to drive definition of mission and scenarios. Just as biological systems are leveraged, ARL should take advantage of knowledge about human cognition in perception and intelligence applications. In the longer term, ARL should use more real robots and should consider testing at Fort Benning with intended user groups.

---

[1]Pew, R. W. and Mavor, A. S. (Eds.). 2007. *Human-System Integration in the System Development Process: A New Look.* Washington, D.C.: The National Academies Press.

HRI needs to be part of MAST research and design. There is good synergy between ARL and university researchers, but there should be more interaction, at ARL and within the CTAs, with HRI researchers in other military laboratories (e.g., the Naval Research Laboratory and the Office of Naval Research, Army Research Office) and academia. To better inform the researchers of ongoing work and needs, ARL should conduct more outreach to invite instruction from academic and military experts and to leverage their work.

The long-term vision is for soldiers to have robotic teammates. In the mid-term, robots could be used as tools, with functions and tasks allocated according to supporting analyses of human and robot capabilities.

Overall Technical Quality

Work on robot reliability and on dynamics of human-robot teaming and RIVET[2] simulation software is notable. The ARL programs have embraced the noble goal of utilizing robots as trusted team members rather than as tools, which is even more challenging than the goal of fully autonomous robotic operation. ARL's Human Research and Engineering Directorate should be involved with and given early priority in all robotics programs. HRI individuals should collaborate with non-HRI groups as horizontal integrators.

Manipulation and Mobility

Advancements and Accomplishments

The research portfolio includes devices at the macro-scale (Big Dog), mesoscale (Rex), and microscales (micro flyer). With respect to legged locomotion, Big Dog and Rex represent the state of the art; they are excellent examples of legged robotic systems. Legs represent capabilities that wheeled and track vehicles cannot deliver; adaptability and robustness to different terrains and scenarios are possible. This is a most appropriate undertaking for ARL. In the case of Rex, theoretical considerations were effectively combined with pragmatic issues in mobility—for example, the ability of a machine to climb a hill or a flight of stairs and the inherent dynamic stability of the platforms. Big Dog is robust with respect to disturbances, and its legs allow for coverage of complex terrain.

ARL is performing high-caliber work on the inherently stealthy micro-flyers. The component-level research in terms of MEMS fabrication, combining actuators with structural functionality, is pushing the state of the art. This system is potentially an organic asset for soldiers for the last 100 meters. Modeling of the flyers was performed to develop ideas about power required, but the details of those models were not provided during the presentation. In addition to component-level research, system issues were addressed to determine the potential range and payload of such a system. Work on flapping vehicles performed on micro-flyers at the University of Maryland using computational fluid dynamics should be applied to the ARL micro-flyer. The work on controllers for micro-flyers is laudable as well.

The use of commercial off-the-shelf robots is appropriate, because it allows researchers to focus on the other research issues at hand. Use of state-of-the-art PackBots for some applications such as mapping for more realistic environments might be helpful. Analysis, modeling, and experimentation are evident within the program, which is appropriate. For example, modeling of robotic motion, actuation, and platform dynamics needs to be performed to develop an understanding of how power is utilized and delivered in legged robotic applications.

[2]RIVET stands for robotic interactive visualization and exploitation technology.

Another good piece of work is the characterization of novel, highly compliant, dielectric and conductive materials, which have application in isolation or as dielectric elastic actuators and electronic components. Characterization and functionality extends through 100 percent strain.

Opportunities and Challenges

There is an opportunity to redefine locomotion for legged systems to increase their efficiency and robustness. Not enough resources are available to create new platforms, but the platforms under consideration can be improved, bringing these robots closer to utilization by the warfighter.

The portfolio should be examined to identify how projects can contribute to concepts of optimal vehicles or vehicle features needed to meet application scenarios. Understanding the efficiencies of vehicles with respect to mission needs and considering vehicles that have not been traditionally studied (such as snakes or blimps) may result in a new paradigm in vehicle attributes that can be leveraged to surpass the defined technical gaps and hurdles. Specific gaps in power generation, power delivery, and countermeasure development should be included in this assessment.

The researchers should more clearly articulate and justify their work—particularly with regard to scenarios for the Robotics CTA. The locomotion researchers should develop joint experiments with other researchers in the ARL, such as the HRI researchers. HRI's vision could be better integrated into the manipulation and mobility projects. Some problems such as sorting of parts, path planning, and robotic arms are not well integrated into any scenarios or even into the Army mission. The issue of reliability should be quantified and captured as part of the research for these platforms.

Systems analysis should be performed for more platforms for various scenarios and should consider energy source, power utilization, robotic counter measures, and payloads. This does not mean that ARL should assume responsibility for research to counter measures, but it raises the issue of the robustness of platforms. Fluid flow analysis should be applied to the micro flyers in order to inform understanding of how to control them. There is a lot to learn about these machines and their performance, and so the interplay of experiment and modeling should continue.

Power for Robotic Vehicles More work should focus on power for the robotic motion, specifically power source (engines and energy storage) and power delivery (actuation, rotation, and thrust). Because of its great advantages in energy density, power density, and the availability of JP-8 fuel, the use of combustion should be studied. New concepts could lead to smaller combustion engines useful for small robots and small unmanned aerial vehicles. In addition, smaller combustors on larger robots could lead to designs with distributed power throughout the robot, which should offer some efficiency advantage.

The area of energy and power utilization presents opportunities for huge advancements. Humans and large dogs use about 100 watts of power on average. A horse is at roughly one order of magnitude greater in its average energy usage at 1 kilowatt, and an elephant uses about 10 kilowatts on average. Why is it necessary, then, for the DARPA/Boston Dynamics Big Dog Robot to require a 15-hp internal combustion engine, which had a little more power at its maximum than an elephant? The issue is that the animal, bird, or human can store energy in a chemical form that is distributed throughout the body near muscles and readily convertible to mechanical energy when needed. Compared to the Big Dog Robot, which uses a hydraulic actuation system, the animal has the equivalent of stored energy in accumulators distributed throughout the body. So, the animal can produce a burst of power without need of a huge engine, while today's robot does not have the accumulator system and has to speed up its huge engine to get the needed power. A robot with a smaller engine running at nearly constant speed and with a well-engineered system of distribution and storage of energy would be much more efficient. Distribution of several smaller engines would also be superior.

To address the need for small engines that run on JP-8 fuel, ARL should review the recent successes at the Rochester Institute of Technology in the development of spark-ignited two-stroke heavy fuel engines.

Flapping Flight The experimental work being done on the flapping flight of nano-scale air vehicles is exceptional. However, the use of rotors and fans should also be investigated. Although some early research indicated that flapping flight may be more efficient than steady rotary motion at nano-scales, recent research does not support this idea. The power for steady motion of a rotor requires overcoming the induced drag and profile drag of the rotor. The power for flapping flight requires power for overcoming induced drag, profile drag, and the inertial power required to stop and reverse the flapping motion.

Flapping may not be the best use of the limited power available on nano-scale air vehicles. Comparing flapping to rotary wing flight might be an appropriate research task that probably should be done, but it might be easier to devote a similar effort to developing rotary winged nano-scale air vehicles. Current work on the quad fan and maple seed concepts may provide a starting point for rotary wing nano-scale air vehicles.

Another opportunity is the engineering of flapping systems; this involves not only their construction and testing, but also how to engineer these platforms. Because they are not fixed-wing aircraft, there is no coherent way to design these systems with respect to power, stability, and controls.

Palm-Sized Aerial and Ground Platforms The objective of this project is to provide the fundamental aeromechanics and ambulatory tools to enhance MAST objectives. The research involves very significant multi-university partnerships, with funding from multiple governmental agencies. Physics related to the vehicles are very different because of low Re and the dominance of the viscous effects. Traditional computational fluid dynamics codes do not have good predictive capabilities, and experimental results are not easy to replicate for the length scales under consideration. Flight vehicles are also highly susceptible to atmospheric disturbances and, therefore, call for new approaches to flight control. The approach consists of experimentation with concepts that loosely model winged flight in nature (e.g., small birds and insects). The focus is clearly on flapping-wing flight, although the lead investigator's background in rotary wing flight has introduced innovative micro-air vehicles such as the quad rotor and the ducted fan rotor. The mass of the vehicles considered in the work ranges from 12 g to 100 g. A parallel effort to understand the mechanics of small ambulatory ground vehicles is less developed.

The work's promise lies in developing a deep understanding of flapping-wing aerodynamics and a much better appreciation of the engineering scaling laws that apply in this environment. The progress in this regard can at best be characterized as limited. The computational fluid dynamics models have shown some promise in qualitatively characterizing the experimentally observed flow patterns for hovering flight. Simple flat plate models have been developed to study the highly coupled aero-structural behavior; a flapping-wing rig has also been built to do fundamental flow measurements to augment the simplistic computational models. Similarly, interaction of flows between two flapping wings has been visualized in the tunnel. If successful, there is considerable value in the work to push the envelope on developing micro-air vehicles. To date, however, the success has been limited, and such micro aerial platforms will develop largely through a build-and-test approach with limited understanding of the physics. The flight of the cyclocopter is an example of such an approach. The principal impediments to microscale ground vehicle ambulation are not well understood. The current approach emulates the mechanics of motion of small insects; designing mechanisms at this length scale may be the goal of this work. Design of both air and ground vehicles at this scale should consider power needs, and very little about this important consideration was presented.

ARL Millimeter-Scale Aerial Platforms This project's objective is to provide the Army with a low-cost, low-observable, mobile sensor platform using PZT and MEMS technology. The study is focused on a feasibility analysis of millimeter-scale robotics. Key challenges in this work include developing and integrating appropriate power sources at this length scale, providing adequate load-bearing capacity for flight vehicles designed using MEMS production techniques, and appropriate framing of the mobility design problem. The focus resides in studying flapping-wing propulsive devices to look at aerial vehicles ranging in mass from tens to hundreds of mg. There is significant dependence on the promise of thin-film battery technology for delivering the power needs for this project.

The approach completely embraces the idea of build-and-fly test concepts, with a focus on MEMS production techniques. Very simplistic aerodynamic theories are used to predict the flight characteristics of these devices, and further improvement is promised as other portions of MAST provide the necessary tools for more refined analysis. To date, the work has resulted in millimeter-scale flapping-wing designs that have been demonstrated on static test rigs. At this length scale, this work represents the state of the art and is one of the most promising developments at ARL in this area. A significant amount of formal design work (including coupled multidisciplinary analysis and optimization, albeit with simplified analysis modules) has been performed in support of this project. Understandably, the work has been well received in the literature, as evidenced by both conference and archive journal papers. The lack of more robust aerodynamic calculations should not hold back this development. The requirements of mechanical power, however, could be a major impediment to success. Parallel developments at appropriate power electronics to get the voltages required for appropriate deformations in pitch and flap should be a continuing focus in this regard.

Overall Technical Quality

ARL is to be complimented for the range of robotics sizes and the different types of mobility devices in its robot research portfolio. Also, the research portfolio is well balanced in terms of analysis and physics-based modeling and experiments. It addresses real-world effects and has great focus on meeting a specific need. Metrics are fairly well defined, and the inherent requirement of a test vehicle drives the system thinking and approach. The work on the "micro flyers" and legged robotic systems are best-in-class. There was a good portfolio of vehicles and platforms from small scale to meso- and microscale. Specific efforts were judged to be leading the state of the art in their areas. Briefers are aware of and addressed the system perspective. A candidate area for increased emphasis is discrete awareness of mission, sensors, and power requirements to meet the application vision and scenario. The efficiency of existing robotic systems in transferring energy from the engine to the environment is still several orders of magnitude worse than biological devices, and therefore continued work in this area is required. Because of the burden imposed on soldiers by battery packs and robots' limited time on mission, research that improves the overall energy density and the efficiency of converting energy to motion is required. In particular, research on the combustion of small JP-8 combustion engines and research on the efficient creation and transfer of force to the environment should be added to the research portfolio.

7

Weapons and Materials Research Directorate

INTRODUCTION

The Army Research Laboratory's (ARL's) Weapons and Materials Research Directorate (WMRD) was reviewed by the Panel on Armor and Armaments at Aberdeen Proving Ground, Maryland, during June 14-16, 2011, and May 7-9, 2012. The theme of the 2011 review was warfighter protection; the 2012 review was focused on lethality research and development (R&D).

The Army Research Laboratory is the corporate laboratory underpinning the operational commands for the U.S. Army, and its WMRD serves as the locus of the fundamental science and technology (S&T) research on materials issues supporting warfighter protection and lethality. The presentations to the reviewing panel outlined the breadth and scope of WMRD's research efforts during 2011-2012, which span the gap between basic research that improves the understanding of scientific phenomena and technology generation that supports weapon and protection system developments and fielded system upgrades. The directorate executes its mission of leading the Army's research and technology program to enhance the protection and lethality of the individual soldier and advanced weapon systems.

CHANGES SINCE THE PREVIOUS REVIEW

WMRD's focus has significantly changed. The quality of its research is greatly improved, and its interactions with other laboratories, both governmental and academic, have developed into active collaborations. WMRD is to be commended for these improvements. For example, several past reviews observed that computation did not often play a key role in research accomplishment. Now there is a clear commitment to transforming materials computation into a reliable predictive tool, as evidenced by a new thrust toward multiscale modeling and toward model validation, verification, and uncertainty quantification.

The WMRD armor technology program has undergone a major change in focus during the past 2 years. At the time of the last review (2009) the WMRD armor program was primarily focused on developing and demonstrating armor technologies that could meet Army Future Combat Systems (FCS) requirements and on meeting urgent Army needs for upgraded combat vehicle armor for use in Iraq and Afghanistan. Since then, the Army has cancelled the FCS program, announced plans to initiate a new Ground Combat Vehicle (GCV) development and acquisition program, and fielded a number of combat vehicles with improved protection that employ WMRD-developed armor technology. These developments have provided WMRD with an opportunity to restructure its armor S&T program to better meet long-term Army protection needs. WMRD has taken advantage of this opportunity by developing plans to support GCV armor requirements and by following a new back-to-basics approach that has led to the establishment of longer-term, ARL grand challenges in protection. The restructuring of the WMRD armor technology program will help the Army to ensure that new and improved armor technologies are available to address all currently projected and future emerging threats.

WMRD has streamlined its core armaments technology programs to make them consistent with the Army's current "squad-centric" focus aimed at increasing the combat effectiveness of small units. New efforts have been established in four areas: (1) low-cost, hyper-accurate munitions technologies; (2) disruptive energetics and propulsion technologies (disruptive technologies are those that provide significant improvements over current technologies, often through sudden innovation); (3) lethal and scalable effects technologies; and (4) soldier lethality technologies. Efforts in advanced propulsion technology now place more emphasis on small caliber munitions and less emphasis on artillery and tank cannon propulsion. WMRD has established a solid armaments technology program that will likely be capable of providing improved technologies to meet future Army needs for advanced weapon systems.

WMRD technical staff continue to show tremendous enthusiasm for their work and awareness of the importance of the solutions being generated through research. They have universally embraced the need to marry experimental and empirical protocols, long established at ARL, with the emerging multiple-scale modeling efforts. Early-career staff lead these efforts, aided by mentoring from the more senior staff. Likewise, the practical aspects of the ARL mission are repeated everywhere, and it is important that actual military vehicles and weapons remain a part of the underpinnings. A passion for timely solutions is well demonstrated, which provides urgency to the research. WMRD's efforts address fundamental and pressing Army needs, and so any means to accelerate progress is of great national importance—a message that is clearly embedded in the program's scientific staff. Networking is evident, but fostering of relationships with outside speakers and visitors would be beneficial. External reviews and emerging new university linkages are encouraged.

ACCOMPLISHMENTS AND ADVANCEMENTS

The connection between ARL and the Army Research Office (ARO) is working well. ARO staff have developed strong connections with ARL personnel and programs and have cleverly incorporated ARL interests into the ARO-sponsored programs. A number of recent examples of links between ARO-sponsored projects and ARL interests were described, including projects that involve exchange of personnel (in both directions), use of facilities, transfer of expertise, ARO-sponsored students who became staff at ARL, and seminars and workshops on various topics. In addition, there have been important examples of technology transfer from the ARO-sponsored projects to ARL programs.

The researchers' excitement for their work is contagious, and investments in human talent of the past several years have produced exceptional results. Interactions with universities in the region have resulted in long-term relationships with appropriate faculty and advanced degrees for ARL personnel. WMRD staff has revealed a strong and healthy environment for creative interdisciplinary interactions.

Protection

WMRD's progress over the past 2 years in the development and integration of modeling and simulation is impressive. In several cases, the integration of simulation and experiment has yielded results not possible with either of these approaches alone. The 2011-2012 reviews revealed a new commitment to transforming materials science computation into reliable predictive tools, as evidenced by new thrusts for multiscale model development, validation, and verification, and for capturing variability through uncertainty quantification. These developments are reminiscent of the Department of Energy's maturation of computation for the purpose of stockpile stewardship, the results of which have been transformative.

There is evidence of quality research programs that allow "out-of-the-box" investigations leading to new discovery, or better yet, to applications beyond initial intent. For example, the study to characterize ceramic microstructure via capacitance measurement and then to correlate to ballistic measures is one clear example of out-of-the-box thinking, because it recognizes the statistical characteristics of the measurements and cleverly uses Bayesian methods to extract dominant ballistic behavior.

Efforts in materials chemistry reflect high-quality work spanning the range of science to engineering. Combinatorial chemistry and property modeling of the behavior of individual molecules uses state-of-the-art chemical selection as pioneered by the pharmaceutical industry, where there are hundreds of millions of possible molecules to synthesize and combinatorial chemistry down-selects a small number of best candidates for experimental trials. At the other extreme, off-the-shelf materials with careful selection of ligands and surface activity are used to develop cost-effective coatings that can be adapted rapidly to a wide range of specific threats. In the future, the merging of the combinatorial chemistry and science with the surface engineering of coatings should provide a strong foundation for future important contributions.

Furthermore, fundamental studies of polymer networks and tunable microstructures, combined with use of dynamic mechanical testing, diffusion properties, mechanical properties, and morphology studies to understand these crystalline and amorphous microstructures, will add value to the entire materials knowledge base.

Many of the armor technology efforts reviewed were impressive. For example, the kinetic energy (KE) armor technology effort illustrates how a back-to-basics approach can potentially provide a significant, long-term payoff for the Army. Some of the initial WMRD KE armor research employed very high quality, small-scale, reverse ballistic experiments. The early experimental results led to the unexpected discovery of a new KE projectile defeat mechanism. During the initial phases of this research, it was not clear that this new defeat mechanism could be scaled up and used in full-scale designs of practical armors. The experimental results were combined with multi-dimensional computational modeling to better understand how this defeat mechanism worked and how the armor technology could potentially be improved and/or optimized. This effort eventually led to a multi-year, applied R&D effort focused on further maturing and demonstrating this new KE armor technology in practical armor designs. This WMRD work has recently resulted in full-scale prototype KE armors that show significant tactical potential for use in GCV and other applications. All of these developments illustrate how a high-quality, well-thought-out experimental and computational modeling approach can improve understanding and thus help meet future Army protection requirements.

SiC tiles for passive ballistics protection have been developed through an integrated experimental and statistical analysis project. Recent experimental advances have provided the capability to measure a broad variety of tile properties via dielectric spectroscopy, leading to detailed spatial profiles of their electrical and mechanical properties for large numbers of tiles. Current work now focuses on the connection of these properties to ballistics measurements. Data analysis is based on information theory, genetic algorithms, and Bayesian analysis, with results suggesting that combining these measurements can lead to a successful correlation. These results may change tile specifications and thereby lead to more uniform and effective passive protection components.

A practical route for the fabrication of large-area panels of nanostructured AZ31-Mg alloys, suitable for lightweight armor applications, has been demonstrated. Specifically, using equal channel angular extrusion, it has been shown that the grain size of polycrystalline material can be reduced to 300-500 nm, doubling the tensile strength while maintaining 15 percent elongation. It is anticipated that further reductions in grain size to the 10-20 nm level could yield even greater improvements in mechanical properties, and thereby armor performance. A similar development is expected in Al and its alloys.

The polymer work is exemplified by well-defined project objectives, well aligned with WMRD objectives. The projects extend process-structure-property relationship studies to new regimes—for example, very high strain rates, balance of functional and mechanical properties by control of chemistry and processing, and definition of three-dimensional (3D) damage to high and low modulus polymers by well-defined projectile physics and impact characteristics (such as size and speed). The close integration of experimental and computational techniques within project teams is synergistic and positively affecting progress.

The engineering science work is impressive. A variety of diagnostics have been used to characterize fundamental armor mechanisms. Flash x-ray, proton radiography (at Los Alamos National Laboratory), polarized imaging, shadowgraph (back lighting), and other techniques are used. The WMRD experiments are complex and have been conducted quite well. The researchers were enthusiastic and spoke competently about their techniques.

All of the programs embrace modeling. The integration over multiple lengths or timescales via efforts to simulate bulk properties linked to atomistic-scale attributes is a very challenging topic in materials engineering. The research team is making progress, and the current efforts are likely to mature in the next few years.

In the mechanics of materials research there is a strong effort to advance transparent armor. The research team is very enthusiastic, mission-focused, and rightfully proud of its contributions. The team is more than adequate for the tasks, with an excellent marriage of a range of disciplines. In the work, the effects of composition on toughness are linked to the atomistic defect structure, and in turn processing parameters are evaluated in light of the desired properties. A key development is in the impurities that facilitate the reaction and sintering with microstructure control. This work reflects an impressive combination of modeling, processing, and examination of linkages to performance and uses both modeling and practical experiments. Although details of the models were not addressed in the short discussions, still the overview seemed to convey sophistication and an effective team. No information was provided on the specific modeling tools, software, hardware, or needs, so it is assumed the combination is more than adequate and properly supported. Integration with external suppliers was evident, but no comment was offered on the relationship to possible system vendors.

Penetration of armor is a fundamental issue for the Army, leading to much research on high-strain-rate fracture. A large body of testing is required to construct accurate models, which is a significant barrier in terms of number of samples, testing time, and expense required to hone the predictions. Attention to minimum testing to validate and refine existing models is required to expedite future efforts. The balance between experiment and statistical analysis is appropriate. The research on penetration is well aligned to critical issues. Some comparable testing is being performed in other federal laboratories; for example, the National Institute of Standards and Technology has considerable prior work in prediction of failure for brittle systems (ceramics) that might be relevant.

Research on lightweight aluminum alloys is important and has directed much attention to how technical gains can be propagated quickly to the field. Features such as grain size or differences in grain boundary segregation or internal porosity need to be understood. Scale-up issues are inherently part of this research and will eventually be addressed. For example, the relative merits of a monolithic cast-forged structure versus assembled pieces (most likely by welding) will eventually need to be considered with respect to the possible vendor base (apparently limited at this time), property differences, and overall unit cost. The team is qualified and working in an area where there are essentially no peers. There are external vendors to advance the work, so there is no apparent limitation in terms of internal capabilities. It is unclear whether the team is properly linking early with existing manufacturability capabilities.

Other examples of high-quality research endeavors include the following:

- *Textile composite armor.* The multiscale modeling has reached a mature level and provides not only answers to 30-year-long performance questions about protective textile composite materials, but also a framework for successful design of lightweight textile armor of the future. The work is both systematic and unparalleled outside of ARL.
- *Aluminum oxy-nitride*. The senior ceramics personnel are mentoring researchers in several areas of simulation; they meet, discuss, and exchange progress. Although AlON is a commercial product, this research may produce unexpected benefits in determining more suitable means to synthesize, consolidate, or otherwise fabricate the material. At the same time, there could be tremendous gains if new compositions or microstructures are discovered to increase the plasticity of the system. Pressure-induced phase transformation can yield nano-scale grain size.
- *Modeling high-strain rate behavior of glasses and crystal plasticity.* Formalisms combined with computational implementation for single crystal and multi-grain forms are well conceived and executed. Care is taken to explicitly account for high-pressure states involved in impact conditions.
- *Armor physics*. The experimental work on armor physics and testing is outstanding as a whole. Within this work two efforts stand out as being of especially high quality, both concerning diagnostics of jets. The jet temperature measurement technique is unexpected in its simplicity, and it has already shown value in validation of plasticity models. Likewise, the imaging is providing important new insight into penetration and mitigation physics that will be vital for both design and validation.
- *Layered nano-materials manufacturing* capitalizes on in-house expertise in larger area processing of polymer sheets—for example, fabrication of polymer-supported carbon nano-tubes, grapheme sheets, and even nano-diamond film—as a means to provide reinforcement as protection materials.
- *The polymer research programs* for the mission that ARL is undertaking are very good. The experimental and computational collaboration within the polymer research programs is outstanding. The fundamental understanding of the structure-property-processing relationship for the polymer systems being studied in polymer networks and tunable microstructure properties is excellent. The combined use of dynamic mechanical testing, diffusion properties, mechanical properties, and morphology studies to understand the crystalline and amorphous microstructures will add value to the entire materials understanding.

Lethality

There is a developing continuum of work, coupled to Army-specific needs, from basic research through applications. Real barrier problems are being approached with the fundamental science necessary to achieve revolutionary advances. One notable example is the work in new forms of energetic materials. The development of new high-energy-density materials has led WMRD to renew its interest in ballistics phenomena with a clear commitment to address issues of both internal and external ballistics. The specific work on extended solids and nano-diamonds for structural bond energy release is technically high risk and potentially high payoff. The 885A1 system successes in lethality and in reduction of environmentally sensitive materials are particularly noteworthy. Deployment of the system without complete laboratory validation has proven to be a wise and timely choice.

WMRD deserves significant credit for its support of dedicated facilities for advanced energetics development. This was an investment of several millions of dollars, and although the various aspects are not unique, the combined facilities form a capability to synthesize, characterize, formulate, and test new materials in a state-of-the-art facility. Other laboratories are moving away from this technology capability that is critical for the warfighter. In addition, the work on extended solids and nano-diamonds for structural bond energy release is technically high risk and potentially high payoff. These experimental efforts are supported by computational work. The extended solids work builds on work at the Lawrence Livermore National Laboratory and is motivated by a project of the Defense Advanced Research Projects Agency (DARPA). The accomplishments to date are encouraging, and if the work can be scaled for full characterization, then the results will be very interesting.

The cold spray capabilities continue to show great promise for applications. This work has demonstrated that reactive materials can be used to tailor the lethality profile of the fragments as a function of distance with a density high enough to compete with conventional inert fragments (7 g/cc). It was found that processing challenges exist in making homogeneous materials (i.e., density gradients) using cold spray. Would composite particles (pre-mixed by milling) result in more homogeneous samples, as well as in tunable reactivity?

The development of low-cost hyper-accurate munitions is one of WMRD's grand challenges. The purpose of this program is to develop precision munitions that will operate in a global positioning system (GPS)-denied environment with the use of flight control algorithms for mortars and artillery rounds by understanding and utilizing the aero-mechanics. Current ballistic rounds are costly to replace (about $1,000 per round) but are relatively low performance compared to an Excalibur-type round, which is a high-performance precision munition. Achieving the Excalibur-type performance with a target cost of $10,000 per round is the goal of the Very Affordable Precision Projectile research program.

The high-pressure polymerization of CO and N is very interesting and cutting-edge work in the polymer field. In this work it is important to characterize Poly CO and Poly N as polymers with their properties such as molecular weight and glass transition temperature. The goal of producing 1 gram of polymer this year may not represent enough to characterize these polymers. These characterizations are important for validating the computational data on the various species being predicted.

Other examples of work of good technical quality include the following:

- The application of commercial off-the-shelf microelectronics for guidance, navigation, and control (GNC) systems has resulted in development of these competencies within WMRD. The addition of on-board intelligence in GNC to achieve accuracy at acceptable cost is noteworthy.
- The crosscutting study of small caliber projectile design, which correlates design parameters with projectile damage and then further correlates damage with the ability to incapacitate an enemy, has led to a sophisticated understanding of projectile effectiveness and has translated

into the adoption of more effective ammunition for use by combat troops. Overall, this work represents an impressive success.

- The polymer work on mechanical properties of soft materials demonstrates a high-level understanding of the issues.
- Low-cost hyper-accurate munitions technology offers considerable potential, and is greatly needed by the Army. WMRD's technical approach is very sound.
- The research on technology to improve the performance of small arms against hard and soft targets is clearly aligned with improving capabilities of small units and individual soldiers. WMRD can do a good job is this area, taking advantage of its integrated capabilities and providing a big payoff for soldiers.
- The work in the computational fluid dynamics (CFD) simulation of the combustion and flow processes in the hypergolic pulse engine is very important and useful to the Hellfire missile program. The work is highly commendable. WMRD's approach is organized in logical fashion by considering a detailed chemical kinetic mechanism, developing a reduced mechanism, and performing fluid dynamic simulation of combustion processes of IRFNA and TMEDA-DMAZ impinging jet-induced spray.[1] Further consideration and treatment of the dense spray behavior will help to refine the existing model and solution.
- The work on CFD simulation of interior ballistic and muzzle blast phenomena is commendable. Using the modified NGEN code, the researchers were able to predict the shot start process being initiated by the intragranular stresses transmitted through the compressed aggregate of propellant grains. WMRD has demonstrated its findings for both large (XM829E4) and small (5.56 mm) calibers to industrial companies. WMRD researchers have appropriately extended their interior ballistic calculations to the muzzle blast zone. The suppression of muzzle flash is an important area to address in detail. Collaboration with Banat Laboratories in this research area could accelerate progress.

OPPORTUNITIES AND CHALLENGES

Materials by Design

ARL seeks to develop the capability to design, optimize, and fabricate lightweight protection material systems exhibiting revolutionary performance. The approach is to realize a "materials by design" capability by establishing a new Collaborative Research Alliance (CRA) focused on Materials in Extreme Dynamic Environments (MEDE). The focus of the CRA is to advance the fundamental understanding of materials in high-strain-rate and high-stress regimes. The CRA is intended to create a collaborative environment that enables an alliance of participants from academia, government, and potentially industry and/or nonprofit organizations to advance the state of the art and assist with the transition of research to enhance the performance of protection materials of interest to the Army. With an award ceiling of $89.9 million, this effort is clearly a major undertaking that will likely require a new approach and level of management both within ARL and in the management of the multi-university consortium to ensure success.

---

[1]IRFNA and TMEDA-DMAZ are types of fuels and propellants. IRFNA is inhibited red fuming nitric acid, TMEDA is tetramethyl-1,2-Ethylenediamine, and DMAZ is dimethylaminoethylazide.

The ARL multiscale material modeling program is a very ambitious effort. It relies on university partnerships to provide and develop many aspects of multiscale materials science. Excellent partners with strong and appropriate backgrounds in these areas have been identified. This is a noble effort if the university partnerships can be coordinated under ARL leadership. These new university programs, especially MEDE given its large number of participating universities, will require effective management on both the university and WMRD sides of the relationship to ensure success. However, no clear management framework has been identified for this purpose. Several areas are not currently well developed. For example, the methodology to bridge scales (particularly coupling to the continuum level for system-level modeling) is still a research area in the general material science community. Most importantly, the approaches to assess the accuracy of material physics require a deep understanding of validation. Key to that is linking to experimental studies that probe the scale of intent. Continuum-level experiments and techniques are often insufficient for validation at reduced scales. Furthermore, many uncertainties will be encountered in these assessments, and uncertainties propagate across scales. Quantifying these uncertainties is an important aspect of validation. The Department of Energy (DOE) sponsored an Advanced Simulation and Computing Academic Strategic Alliance Program to address these same issues. ARL should review this DOE program, because the university partners are perhaps several years ahead in uncertainty quantification, and their model for university interaction bears close examination. Also, in the effort to incorporate computations, modeling, and simulation in the research, it is important to sustain the commitment to leading-edge experimental methods and validation at all scales. Examples of needed leading-edge facilities include electron microscopy with atomistic resolution; ballistic measurement facilities that provide three-dimensional, digital kinetic information with innovative, state-of-the-art measurements systems; and innovative validation at scales consistent with the multiscale analyses.

"Materials by design" is a grand challenge that pushes the high-performance-computing envelope, and its attainment will require a significant commitment of time and resources. In principle, accurate and efficient multiscale modeling will replace costly and time-consuming experimentation to discover process-structure-property relationships; it proceeds from a set of desired properties (or performance criteria) in a systematic way to identify material structures that display the desired properties. Generally, the systematic aspect of the design process utilizes process-structure-property relationships, which reflect the control a process places on structure and the control that structure places on properties. The current limited ability to design materials stems from an incomplete knowledge of all of the process-structure-property relationships. Furthermore, because design is the inverse of multiscale modeling, ARL's solution to this problem will only be found through a systematic search of materials structure space for desired properties and performance. Solving the inverse problem has taken on a mathematical definition in recent years, although the principle remains the same—to work systematically from properties to structure. In addition, extension of mature multiscale modeling to manufacturing considerations will ensure that materials or designs remain in the window of practical solutions. Much more progress is needed to integrate this competency with manufacturing and affordable solutions. Rather than use phrases such as "multiscale modeling" and "materials by design," the programs should use the phrase "integrated computational materials engineering," while still encouraging multiscale modeling. In this way, the emphasis would shift to materials engineering, which is where ARL's emphasis should be.

The programs in materials and manufacturing science for lethality identified the following current and future themes: materials by design—a multiscale approach; materials in extreme dynamic environments; field effects in materials; emerging materials; life prediction of materials; and enabling agile manufacturing science

During the past few years, WMRD has expanded efforts aimed at materials modeling, as evidenced by the first-principles density functional theory (DFT) analyses on interface properties in rare-earth modified silicon nitride (Si_3N_4) for ceramic armor, and the research on chemical kinetics and reactions paths that provides input for computational fluid dynamics models for rocket propulsion. In addition, crosscutting efforts build on other computational expertise within ARL, notably on survivability with the Survivability/Lethality Analysis Directorate.

A higher level of the WMRD organization has goals for multiscale modeling, materials by design, and integrated computational engineering. A more detailed strategy for the development of a suite of modeling capabilities along with validation and verification plans is needed for continued success in this area. Developing this strategy is important in light of managing the new MEDE program. An approach that has been successful in other DoD-oriented organizations involves the selection of one or two foundational problems that address a system or subsystem (e.g., a medium-caliber weapon) and motivate the development of a suite of complementary computational and experimental tools. WMRD should clearly define the classes of computational tools needed to solve important foundational problems and should articulate a timeline for their development, validation, and application to problems. The role of external organizations, including DOE and universities, should also be considered carefully.

Computational and Experimental Tools

The development of a complementary suite of computational and experimental tools is expected to permit WMRD to respond to the needs of the soldier at the unit level with unprecedented speed. Development of this infrastructure will require new investments in state-of-the-art characterization and testing capabilities, particularly in the areas of microscopy and spectroscopy, in order to validate models across all length scales and to rapidly explore the design space. The work on detonation characterization is an example in which development of small-scale experimental tools, combined with modeling, enables more rapid exploration of novel, insensitive energetic materials. In the area of characterization, strategic partnerships will be critical for success, because no single institution can maintain all classes of advanced instrumentation.

Integration over multiple length or timescales is a very challenging topic in materials engineering and the simulation of bulk properties linked to atomistic-scale attributes. The extension of mature multiple-scale modeling to manufacturing consideration will ensure that materials or designs remain in the window of practical solutions. Little attention has been directed to identifying the needed attributes for the testing or forming equipment. It may be impossible to attain the desired combinations of pressure, temperature, time, strain rate, or other manufacturing attributes. Polymers decompose, metals melt, diffusion occurs, and stress or temperature-induced phase transformations occur, and these boundary conditions have to intersect with tooling constraints (strength, for example), loading rates (press design), and practical limitations. WMRD made repeated mention of cost, and although some of the novel penetration-resistant materials are of great importance, the researchers did mention the inability to lower cost. Extension into the arena of cost is yet another aspect of multiscale modeling. That is, are the materials simply too costly (aluminum alloyed with silver, for example), or is the fabrication window too narrow, incurring an excessive cost? There has been good progress, but still much more is needed to integrate the materials into manufacturing and affordable solutions.

The research program is broadly based on sound conjectures and relies on contemporary tools. The research tools are adequate, but with limited technicians one would expect to see more automation in the experimental tools. However, practical aspects are elusive at times, and the solutions offered sometimes can be at odds with statements about cost considerations. For example, gradient microstructures are commonplace in sintered tungsten carbides, requiring only an atmosphere adjustment during sintering, but the early trials with cemented carbide projectiles lacked appreciation of this dimension. Indeed the early materials were taken from lower quality materials. Partnerships should be formed with leading teams in technologies for which WMRD is not up to date. In the case of the cemented carbides, it is most appropriate that teaming agreements be made with leading firms such as Kennametal. Overall, although partnerships exist for several projects, the critical participation of leading partners is missing. WMRD can benefit from more active outreach to experts and from broader involvement by experts in individual programs. WMRD researchers mentioned that joint publications were discouraged, but such collaboration should be encouraged.

Nanoscale materials can be consolidated by new combinations of temperature-pressure-time. Multiple-scale modeling will identify the combinations of greatest merit, which will likely lead to lower processing temperatures and a need for much higher consolidation pressures. Higher pressure consolidation and shaping of materials (e.g., polymers, metals, ceramics, and composites) is a very desirable new research direction for WMRD. The work should anticipate pressures in the 2 to 5 GPa range and simultaneous temperatures in the 1,400 °C range. Short cycles are of great merit.

Modeling and Simulation Challenges

Because many physicochemical processes, simulated by ARL researchers, involve multiscale, multidimensional, and multiphase flows with heat and/or mass transfer under dynamic conditions, numerous challenges exist in the modeling work as well as in numerical simulation of these processes.

Because of the complex physical and chemical processes described above, scaling laws are expected to be very difficult to develop for armor penetration process. There are multiple pathways for material to respond to its physical contact with the impacting or penetrating projectile and to the combustion-induced environment variations. For example, the material used in the armor plate can go through mechanical fracture process. The instantaneous contact surface(s) between the high-temperature copper jet depend upon many detailed interactions between the hot penetrating materials and the deformed armor plate components. To understand the potential heterogeneous reactions between the gas-phase chemical species and the instantaneous surface of the condensed phase materials, one has to consider the nanoscales in terms of active reaction sites, the microscales for chemical kinetics, the mesoscales for species mass diffusion across the layer with strong concentration gradient, and the integral scale in the processes associated with jet penetration, convective heat transfer in the flow channel, conductive heat-transfer process, and mechanical deformation in the condensed phase. Besides the matter of different reaction paths, the armor material may evaporate and react with the ambient penetrating gas mixture. Some combustion products may also adsorb on the surface. Some metal plates may have molten layer on their surfaces. High-speed flow may introduce Kelvin-Helmholtz instability associated with surface wave breakup and droplet formation. Turbulent multiphase reacting flow also requires consideration of a broad range of length scales, from smallest Kolmogorov length scale to the integral scale. Surface roughness and/or contour variation with time should also be considered, because armor components usually do not consume at high rates. The timescale for their deformation and consumption usually differs from the short timescale for chemical reaction and jet penetration. The influence of protruded edge of damaged armor plates may induce vortex shedding phenomena and therefore influence the mixing process between the fuel-rich and oxidizer-rich species. These are some of the important processes to be considered in the model formulation.

From the numerical simulation point of view, wide ranges of scales are present in the armor penetration processes that differ by orders of magnitude because of both turbulent and multiphase reacting processes. Therefore, massive computational resources are required to reasonably resolve those scales. A large-eddy simulation (LES) approach can be applied in this problem, for which adequate subgrid scale models need to be developed. The direct numerical simulation (DNS) approach is expected to be far from suitable for any of the armor plate penetration simulations at the present time. Because this problem involves liquid-gas interface, liquid-solid interface, and time-varying diffusion flames in the gas phase zone, interface capturing (or tracking) becomes an important requirement to achieve. Thus, a level set technique should be considered for treating time-varying interfaces.

The Disruptive Energetics Program

The "disruptive" energetics program conducts leading-edge research in new material design largely based on state-of-the-art molecular modeling toward developing new energetic materials that surpass the energy densities of current materials. Several WMRD researchers in this area are well recognized in this field. However, the focus on energy density as the metric of performance may be too restrictive. The manufacturability and stability of these meta-stable materials are important aspects to consider for practical application. The desired energetic materials also have to readily convert stored chemical energy almost directly into PdV work (refers to expansion work) rather than into just heat (thermal energy) if it is to be used in fragmenting or blast munitions. Determining these transformation properties usually requires material quantities that are much greater than those associated with the atomistic scales. Although the laboratory uses the traditional experimental methods to assess heat of formation and density, these techniques may be insufficient to determine appropriate energetic material characteristics. To complement the developing simulation capabilities, some research should consider developing experimental diagnostics (i.e., ultrafast laser or line-Visar) that can probe the smaller scales appropriate for assessing the atomistic, subgranular, and mesoscales. The diagnostics may be necessary to determine the bridging of scales (particularly linking to the continuum scale), but it is currently lacking in the research portfolio.

WMRD has initiated a new disruptive energetics technology effort with very aggressive goals for developing new types of high energy density materials (factor of 10+ greater than current materials) suitable for use in Army weapon systems. This is an extremely challenging problem, and the WMRD staff needs to ensure it is taking advantage of all of the related work that has taken place elsewhere, both in the United States and overseas. The goal WMRD has set of being able to provide the power of 155-mm artillery in a significantly smaller volume (i.e., 40-mm-size to 80-mm-size munition) is an extremely challenging problem.

The high risk inherent in extended solids and nano-diamonds as energetic materials can be mitigated by the development of new theoretical and experimental tools that are applicable to energetic materials in general. To this end, WMRD needs to complement the traditional experimental characterization methods by collaborating with other university and national laboratory experimental programs using advanced diagnostics for assessing the atomistic, subgranular, and mesoscales, specifically, the local electrode atom probe at Iowa State University, Colorado School of Mines, and Northwestern University; the ultrafast laser at Lawrence Livermore National Laboratory and Los Alamos National Laboratory; and the hot stage scanning electron microscopy at Lawrence Livermore National Laboratory. Investing in advanced detonation diagnostics such as line-Visar, available at Sandia National Laboratories, would benefit both high- and low-risk efforts (e.g., DEMN—a type of explosive). The nano-diamond work is interesting, but additional work, some of which is planned, is needed to make measurements beyond temperatures. A question to ask is: Would a detonating system be expected to yield measureable effects?

In the study and development of poly-CO, WMRD uses systematic approaches to explore the properties of this new energetic material, whose energy density is higher than those of conventional explosives (e.g., RDX, HMX, and CL-20). The estimated heat of explosion (ΔHexp) of poly-CO is around 8,130 J/g, which is higher than that of RDX (5,700 J/g) and CL-20 (6,344 J/g). In addition, the estimated mass density of this new material is 2.46 g/cc, which is also higher than that of RDX at 1.8 g/cc and CL-20 at 1.96 to 2.04 g/cc. WMRD's systematic approaches include:

- Verifying the stability of the poly-CO using hysteresis characteristics of this material after compressing it to 10 GPa and then dropping the pressure to lower levels;
- Studying its phase diagram and establishing boundaries between delta, alpha, beta, epsilon, and fluid phases;
- Determining its decomposition temperature around 650 K, which is higher than those of RDX and CL-20 at 435 and 468 K, respectively; and
- Studying the deflagration emission enhancement of RDX/nano-diamond mixture from pure RDX, using pulse jet laser with a wavelength of 1.06 μm.

However, this material's effectiveness for propulsion and detonation application is not clear. First, the heat of explosion (ΔHexp) is not as high as the teams believe, around 8 to 10 times that of RDX and/or CL-20. It is only 1.28 × ΔHexp of CL-20. Second, because of the relatively high atomic weight of both carbon and oxygen, the molecular weight (Mw) of the combustion product of poly-CO will not be low enough to become highly attractive. This is based on the fact that both specific impulse (Isp) for rocket propulsion and impetus (Im) for gun propulsion are dependent on the following equations, where Tf refers to the flame temperature:

$$\mathrm{Isp} \sim \sqrt{\frac{\mathrm{Tf}}{\mathrm{Mw}}} \qquad \text{and} \qquad \mathrm{Im} \sim \mathrm{Tf}/\mathrm{Mw}$$

The group should consider evaluating the propulsive performance of poly-CO using the thermochemistry computation with NASA-CEA Code, Blake Code, or Cheetah Code before putting further effort into its development. Third, the amount of material is only available in micrograms, which is too small for any accurate material characterization purpose. A more economic method for its scale-up production in gram quantities would be helpful.

The ARL Sensors and Electronic Devices Directorate (SEDD) has been a major contributor to a DoD and Army multi-year advanced power and energy research and development initiative. The results of this initiative should be reviewed to determine if any new, low-cost technologies for producing and/or storing energy can power novel advanced protection systems and help to improve vehicle and/or soldier protection.

Human Models

Development of improved human models is both challenging and controversial. Not only are there multiple scales, both temporal and spatial, in the modeling of the brain subjected to primary blast, but also it is not clear how to relate the stress-strain history calculated at the continuum level to physiological and cognitive damage. The subscale damage may not be modeled directly. Instead, careful measurements of the damage of insulted biological materials should be correlated with prescribed insult levels. However, the use of continuum modeling techniques of complex structures such as the brain (neuronal networks, highly vascularized) or extremities (hard material connected to soft material by connective tissues of intermediate properties), in a way that the data can be translated to actual injury, is highly ambitious, If successful, however, this work would make an extraordinary contribution to the understanding of such injuries and ultimately to the development of protection strategies.

Collaborations and Other Interactions with the Professional Community

Given the complexity of multi-scaling research, ARL should consider using an interagency approach. Many other government agencies internal and external to DoD are attempting to understand similar problems, and ARL can take advantage of synergies to fast track its computation and computation method programs. ARL can assume the leadership role in this research area by organizing an interagency conference focused on multi-scaling research. Understanding what the laboratories at DOE, other parts of the Army, Air Force, Navy, Federal Aviation Administration, and the academic community have attempted in this area would be very beneficial.

Although the phrase "low cost" was mentioned repeatedly by WMRD staff, no consistent description of cost analysis or cost-benefit tradeoffs was provided, and some of the researchers seem to be moving into very expensive processes to lower material cost. There have been attempts to disrupt prior thinking and to find disruptive solutions. However, it is difficult to be disruptive from an insider position. It will be important to properly construct and empower programs so they can be truly disruptive. Often, those with nothing to lose, that is, outsiders, can be most effective in constructing and empowering programs so they can be truly disruptive, which suggests that parallel organizations, Small Business Innovation Research (SBIR) contracts, or maybe even university teams could take on this role.

The SBIR program is inconsistent with regard to assisting with the research mission. In a few instances SBIR programs provided visible contributions, but overall this seemed to be ignored. Although there was considerable discussion on filling out the tool box for the investigators, there was no mention of maintaining that array of tools. As computational platforms, researchers, and software change, it will be important to frequently revisit the tool box to ensure that the packages are operable and adequate.

The once robust SBIR program has been de-emphasized. Proposal review may be a contributing factor, but this trend should be examined. Similarly, proposal evaluations of the Director's Strategic Initiative (DSI) and the Director's Research Initiative (DRI) programs, which allocate funds for innovative projects, should be streamlined to reduce the burden on the senior members of the staff.

Researchers are not routinely attending scientific meetings to interact with peers and share unclassified work. One example is the APS shock compression meetings, but there are numerous professional meetings where ARL personnel could benefit from exchanges with their peers. The unclassified work could be presented at such meetings. Because effort, time, and funding are expended on basic research, the talented ARL researchers should be encouraged and supported to interact with their peer researchers, to benefit their classified work and the scientific community. Leadership should encourage archival publication of the experimental results, as well as the computational results. The *Journal of the Army, Navy, NASA, and the Air Force* (JANNAF) could be considered if the work is considered limited distribution.

The goal of assembling a world-class research staff in WMRD will require a sustained effort. The recession of 2007 may have contributed to the surge of excellent personnel joining ARL, but as the recovery of the economy progresses and the DoD budget is reduced, there will be a need to retain the human talent pool responsible for the progress in recent years. One suggestion is to develop a program like the Truman postdoctoral fellowship of Sandia. Additionally, the need to develop programs to mature and mentor junior staff is essential. To enhance the recruiting of highly talented staff, ARL should consider offering a special fellowship that may attract exceptional new researchers to ARL. The DOE national laboratories offer special fellowships (Truman Fellowship at the Sandia National Laboratories, Oppenheimer Fellowship at Los Alamos National Laboratory, and Lawrence Fellowship at Lawrence Livermore National Laboratory) that typically bring in approximately 50 applicants. Formal proposals and letters of recommendation are requested and evaluated by a committee of senior scientists. One or two of these fellowships are offered per year. These fellowships have brought to the national laboratories exceptional researchers who often later become permanent staff members. They are offered as a postdoctoral position with equivalent staff salary and with additional funding that supports self-directed innovative research of their own origin. Perhaps ARL can consider developing a similar fellowship program to attract exceptionally talented researchers that will showcase research capabilities at ARL.

OVERALL TECHNICAL QUALITY OF THE WORK

The overall scientific quality of WMRD's work is comparable to that of comparable national laboratories. WMRD should be commended for its investment in advanced energetic material investments. Many other laboratories are generally moving in the opposite direction with respect to investments in energetic material research, unfortunately. Generally, the WMRD programs reflect a broad understanding of the science and engineering underlying their work. There is generally a strong interaction between experiment and simulation. WMRD has appropriate laboratory equipment and numerical codes and models. Generally, the laboratories are well supplied and state of the art. In particular the shock physics laboratory is impressive. Computational tools are used extensively, but use of characterization equipment could be expanded, and in some cases WMRD could adopt additional numerical codes and models from outside ARL. The research staff is well qualified, enthusiastic, and knowledgeable. WMRD's work reflects understanding of the Army's requirement for research or analysis, and in many cases there is a clear focus on the Army's needs—for example, in the development and fielding of the new bullet. WMRD uses an appropriate mix of theory, computation, and experimentation, as exemplified by the interaction of the grain modeling that predicted the dislodging of the bullet by bed compaction and experimental verification. Also, the balance between 6.1 and 6.2 research and the ratio of customer funding to internal funding are appropriate.

The enthusiasm of the WMRD staff appears impressive. The number of WMRD staff members holding advanced degrees, particularly Ph.D.'s, has been steadily growing. The staff's technical qualifications are impressive, reflecting hiring from many universities. The research staff's skill set is diverse, reflecting hiring from many universities. Recent hires show great capability in contemporary simulation tools. However, the knowledge base is missing certain critical components, such as a connection to the nation's research infrastructure and the practical observations that should predate simulations. These limitations are offset by the researchers' enthusiasm. More linkages are needed, and hiring is not coordinated and planned to bring in top talent. As the science and engineering markets recover, ARL will need to be more proactive in its early identification of talent. Because fewer hires will be possible in future years, care should be taken to ensure that top talent is added. The research tools are adequate, but the limited number of technicians makes automation in the experimental tools a necessity.

The WMRD efforts are focused on critical aspects of improved armor and munitions. The improvements in protection and penetrators create a perpetual contest. The lethality of the new ammunition is well demonstrated and requires a responsive escalation in protection. The cadre of early-career technical hires is maturing and sustaining focus on topics with likely impact, and management allows projects to reach maturity. How decisions on project termination are made is unclear, so a more formal case might be appropriate to show the decision points.

The WMRD armor protection technology and prototype combat vehicle armor development are internationally recognized as best-in-class, as evidenced by the half dozen or so allied countries that maintain active collaborations with WMRD in the armor technology area, as well as by Army near-reliance on WMRD-developed combat vehicle armor technologies. Compared with most of the Army, Navy, Air Force, and DOE laboratories performing weapons-related research, as well as a number of weapons laboratories in foreign countries, the integrated capabilities that WMRD has in the armaments and lethality areas are among the best anywhere. WMRD is not the best in everything it does, but its integrated interior, exterior, and terminal ballistics, lethality, materials, experimental, computational modeling, and systems analysis capabilities are very impressive and considered to be among the best worldwide. WMRD continues to conduct high-quality armament-related research that is leading to new weapon system concepts and products offering advanced capabilities that make a significant difference to soldiers.

The quality of WMRD's lethality research also is impressive. The lethality mission is very important to the Army, and WMRD continues to conduct lethality-related research that is leading to weapon system concepts and products that can make a significant difference to the soldiers. WMRD has a good understanding of the Army needs in this area. WMRD has streamlined and focused its core lethality technology to make it more consistent with the Army's current "squad centric" focus aimed at increasing the combat effectiveness of small units. WMRD's solid lethality technology program will be able to deliver new understanding and many new products to the Army over the coming years. WMRD is doing an excellent job.

8

Crosscutting Issues

NEED FOR TECHNICAL MANAGEMENT STABILITY

Army Research Laboratory (ARL) leadership is in transition. At several levels, from the ARL Director through individual directorates, "acting leadership" is the watchword of the day. The hard work and significant accomplishments of the current acting leaders are acknowledged, but ARL personnel clearly desire stability in the management chain; many leadership decisions will be made elsewhere within the Army. It is important to note that instability introduces uncertainty, which in turn introduces risk of inefficiency and misdirection. This uncertainty is most apparent in the Vehicle Technology Directorate, which is transitioning under the Base Realignment and Closure Act, establishing new research agendas, rapidly hiring new personnel, and making key long-term decisions regarding major experimental facilities. The Army should expedite a return to stability of ARL's technical management in the near future.

The hiring issue is not confined to senior management. ARL has been highly successful in recent years in recruiting many bright, early-career scientists and engineers, often in newly developing technical areas. Although these new recruits offer great promise for the future, they are in need of strong technical leadership. Some technical areas are benefiting from seasoned internal leadership, but in a number of newer areas senior technical leadership should be recruited from outside ARL, because ARL continually addresses emerging scientific and technical areas. Acknowledging limited flexibility in the scientific and technical personnel track and issues of hiring freezes, there may be an alternate strategy for addressing this clear need for senior technical guidance: ARL should consider 2- to 3-year appointments through the Intergovernmental Personnel Act (IPA) process, targeted at emergent laboratory efforts and populated by academic scientists in the area of interest who will use their sabbatical leave for this pursuit. Ideally these individuals would be participants in extramural efforts already in place, strengthening the collaborative interaction and providing senior supervision at a critical time in program development.

ARL IMPACT

ARL has released the first volume of the "Research @ ARL" series; this event deserves congratulation. This volume of technical papers, focused on recent advances in Energy and Energetics (available at www.arl.army.mil/ResearchARL), is the first of many planned documents that will help stakeholders to understand the scope and direction of recent accomplishments by ARL's dedicated and talented staff. Addressing this audience through such technical compendia is desirable and praiseworthy, but it should not be viewed as sufficient to address the fundamental questions asked of every research and development (R&D) organization: How would the U.S. Army be different if ARL had not existed over the past 20 years? and Why should we expect future funding to achieve the hoped-for impacts?

ARL is a scientific and technical (S&T) organization. Most of its funding is of the 6.1 and 6.2 types. Often such organizations measure their effectiveness by identifying technology transitions to others who will carry out the development and technology maturation processes. They have to then convince stakeholders of projected future impact of their efforts. However, in attempting to project the future impact of current programs, any S&T organization is hampered by the inevitable fact that it may take several years or even decades before the full impacts of current programs are realized.

Regardless of how the story is formulated, the stakeholders' confidence in the laboratory and its management will be bolstered by evidence that the decision making and processes of the present are comparable to or better than those of the past that led to measurable impacts. To tell this story properly, many organizations have had recourse to retrospectively tracing the consequences of R&D events in recent or distant past. ARL last performed this exercise in 1997,[1] but there is no recent such activity at ARL.

The full impact of research can only be measured after the fact. Near-term impacts (or transitions) require looking back only a bit and can be monitored during most of the R&D effort. Long-term impacts require deeper historical probes and are more likely to be assessed for only a few notable examples. In both cases, organized processes for gathering and analyzing these data require management attention and designated leadership. Developing these data and presenting them in a manner that is useful for the intended audience is a job best left to professionals specializing in such activities. To achieve these goals, the S&T organization benefits from having an appropriately supported historian with sanctioned access as well as internal report requirements organized to ensure the collection of appropriate data and personal recollections.

It is important to emphasize the potential value of impact analysis for management and staff in the ARL itself. Lessons learned from past efforts can be mined and used to provide context for management self-improvement. New hires can readily learn about the organization's successes and failures to accelerate their effective participation in laboratory efforts. Not incidentally, awareness of impact of the laboratory's work can instill a sense of pride and inspiration in all members of the laboratory, with obvious value to all concerned.

ENTERPRISE MANAGEMENT

This section addresses cross-organizational activities within ARL, which may vary in size from as small as scientist-to-scientist interactions across two directorates to as large as multi-directorate enterprise efforts such as the Network Sciences and Autonomous Systems programs. Organizations whose

[1] *The Genealogy of ARL*, ARL-P 360-2, May 1997, ARL, Aberdeen Proving Ground, Maryland. Available at http://www.arl.army.mil/www/pages/516/arl_genealogy.pdf (accessed October 15, 2012).

reward structure is stove-piped within directorates, such as ARL, may experience uncertainty with respect to who is leading the effort, how priorities are set, and who receives credit for accomplishments. On the other hand, the potential value of such collaborations in a technological world that is growing rapidly at the interfaces between disciplines is recognized by all and should justify increased management focus on collaborations.

ARL's increased attention to enterprise R&D efforts is commendable. As the technical quality and depth within directorates continue to improve, ARL should continue to increase its focus on broad multidisciplinary issues that can only be addressed by collaborative work across several directorates and with extramural partnerships that enhance the ARL intramural capability. During the review period covered by this report, the ARL Technical Assessment Board (ARLTAB) had occasion to conduct separate reviews of the Network Sciences and the Autonomous Systems Enterprises (discussed in detail in Chapters 2 and 6, respectively). In addition, several members of the ARLTAB and its panels reviewed the development of plans for a third enterprise area, multiscale modeling.

These enterprises demonstrate strong leadership by ARL senior management in identifying technical strengths in various parts of ARL and stitching them together under unifying labels. These independently developed areas have the potential for synergism and collective development. In each case there is great technical breadth and depth. However, each enterprise area is in need of more decisive top-down management direction that can encourage and exploit the synergism. The following paragraphs offer several specific cautions and suggestions for management attention in such enterprise management.

ARL management is organized by directorates. Because no program is fully matrix-managed with promotion and salary decisions delegated to an enterprise manager, leadership for the enterprise's activities is spread out among the several directorates, and often such arrangements can blur the leadership structure. Many types of organizational structures can be applied to this situation, but clearly defined leadership by management is required. Without clear leadership, there can be no commonly understood cross-organizational plan, and without such a strategic plan there can be no metrics or goals for management (or external advisory committees such as the ARLTAB) to use in judging enterprise accomplishments.

Each enterprise anticipates major contribution from extramural entities. Without a clear enterprise strategic plan, extramural partners cannot optimally direct their efforts, and there can be no defined template by which ARL managers can judge the importance and impact of these extramural efforts. (A more complete discussion of ARL extramural programs is presented below).

Enterprise efforts evolve in time, resulting in changing priorities that demand flexibility to move resources from low- to high-priority areas, in some instances from one directorate to another. Line-item budget inflexibility may inhibit ARL management in this regard. Sufficient flexibility has allowed senior management to accrue funds in support of Director's reserves and initiatives. These efforts deserve continued support and praise, and they suggest that ARL may indeed be able to take appropriate actions to move resources between enterprise topic areas when required.

Communication is critical to an R&D enterprise. Leadership and technical personnel at all levels should have frequent formal and informal opportunities to meet, share information, and build collaborative programs. Some of the presenters at the ARLTAB reviews seemed to be hearing from their colleagues for the first time. Although this may be applauded as added value from the ARLTAB review, it discloses the need for increased management attention to improving lines of communication. Of course, communication with technical peers from other organizations should also be encouraged through increased travel to and participation in relevant technical meetings, including international professional meetings, and through special workshops organized by ARL to achieve such goals.

Enterprise efforts are very important, but management should be wary of diluting disciplinary excellence in other technical areas as each of the participating directorate's attempts to demonstrate its strong participation in the enterprise.

EXTRAMURAL COLLABORATIVE ALLIANCES

Since its inception as the singular S&T laboratory for the Army, ARL has used several funding and organizational tools to engage a large extramural cadre of scientists in universities, industries, and other government laboratories. These efforts at their best extend the range of expertise focused on Army warfighter needs far beyond the capability that can be localized within ARL. The new multiscale modeling Cooperative Research Agreement (CRA) is the most recent example of this approach, involving three directorates: the Sensors and Electronic Devices Directorate (SEDD), Weapons and Materials Research Directorate (WMRD), and Computational and Information Sciences Directorate (CISD). Other collaborative alliances are the Network Science International Technology Alliance (ITA) and multiple Collaborative Technology Alliance (CTA) programs focused on robotics, network sciences, cognition and neuro-ergonomics, and micro autonomous systems and technology.

Benefits of Extramural Collaborative Programs

ARL is to be commended for the utilization of these extramural programs, which have enabled ARL to establish contact with and provide vehicles for leveraging world-class research in the identified areas. These collaborative alliances are unique in the history of all government laboratories, and they help ARL to bring all this talent to bear on the targeted problem areas. Another real benefit is the sense of community and excitement about the selected areas felt by many bright people within ARL who become involved in the programs and in collaborative ventures with extramural partners.

However, issues arise from efforts to make optimal use of these alliances. It is important to note that neither the ARLTAB nor any of its panels carried out a comprehensive review of any of these collaborative alliancces during this review cycle. Rather, fragmentary information was gathered during the enterprise reviews of Network Sciences and of Autonomous Systems programs and during regular panel reviews of several of the directorates.

Internal Management

A major strength of the collaborative alliance approach, that is, local management by the extramural collaborators, may also be viewed as the concept's inherent weakness. Too much flexibility coupled with too little program direction from ARL management could lead to an environment in which excellent science is done but the benefit to the soldier through ARL and subsequent development programs is not achieved. The challenge for ARL is to evolve the way it monitors and manages its portfolio to leverage the best outcomes from these efforts.

The sheer size and diversity of these large crosscutting projects can often make them appear disorganized to the outside observer. It is also the case that early on in such programs uncertainty about the optimal choices in a large solution space can be best addressed by a shotgun approach that aims at multiple programmatic targets without an integrating plan. Nevertheless, problem size does not preclude a concise visionary statement of the desired outcome, in particular one that includes a clear statement about desired impact on a soldier in the battlefield. This type of vision statement will guard against the temptation to try to address everything from day one and will, instead, expose clear intermediate goals and benchmarks that are imperative to successful management of programs on this scale. Clarity of vision also has inherent benefits for participants, who can more readily tailor their own contributions to achieve the vision.

As the program matures, it should increasingly be managed with a systems approach that clarifies goals, sets metrics, and down-selects from among alternate approaches. Ultimately the responsibility for this approach lies with ARL, but it should be implemented in concert with the academic and/or industry members of the alliance. For some programs, the plan may lend itself to the use of target capability concepts. For example, Micro Autonomous Systems and Technology (MAST) program goals as currently stated have elements of a complete capability concept. Conceptually, the MAST robot is hand-held size and smaller, primarily utilized within distances of approximately 100 meters of the soldier or small unit, and intended to provide increased rapid and mobile situational awareness in complex real-world urban and other environments. To complete the MAST capability concept, ARL management might add metrics for time to complete the mission; reliability (in terms of getting there and back without break-down and of finding the enemy and munitions after completing the mission); and the mission readiness of the robot.

Links to Intramural Science and Technology

ARL should give serious attention to its model for optimizing the effectiveness of alliances in which the partnering organizations are universities. There are two models for ARL to consider, and they are not mutually exclusive. In the first model, ARL researchers are scientific peers in the research collaboration with partners who should be, and often are, academic scientists of substantial international stature. In the second model, ARL researchers perform the invaluable function of translating Army requirements and educating the academic partners of the field constraints, and thereby help transform academic problems into equally challenging problems that also are of interest to the Army—that is, they act as links in the feed-forward, feed-back loops between academia and the laboratory.

Both models have value for ARL, and both have demanding requirements. For the first, collaborating researchers from ARL need to have as deep a fundamental understanding of the science as the university collaborators. To have peer status, the researchers' accomplishments need to be recognized. For instance, the publication records and awards need to be comparable. In the second model, for the constraints from the field to be persuasive and for clever inclusive problem formulations, the intellectual depth again needs to be comparable on both sides.

It is important to reiterate that the ARLTAB has not performed a comprehensive review of any of these alliances, but has drawn impressions from listening to the references to these made in directorate reviews and more complete overviews in the enterprise reviews of Network Sciences and Autonomous Systems CTAs. It appears that both models are in play, but the primary mode of interaction is that of model two, ensuring translation of the academic research to achieve Army impact.

ARL should continue to take steps to enhance the utilization of model one. This opportunity to build the capability of ARL staff through strong direct one-on-one collaborative research with the best academic scientists should not be wasted. To achieve this goal, ARL should adapt the research culture in the laboratory to better encourage, support, and reward collaborative projects between scientific staff and their alliance peers in academia. With sufficient support from the Army, ARL should explore personnel exchanges between the institutions, the co-advising of graduate students and the co-locating of some of them in ARL laboratories, the Director's reserve funding that encourages such collaborations, incentives for joint publication, and other mechanisms.

Review of Extramural Collaborative Alliances

The ARL Director has instructed the National Research Council-constituted ARLTAB, with its six panels of technical experts, to carry out reviews of the six ARL Directorates, with the primary focus on intramural technical effort. Instead of being subjected to regular and comprehensive reviews, the work of the external collaborative alliances is considered at the level of detail deemed appropriate during regular directorate reviews, and the focus of such examinations has been on the work of the ARL participants in external collaborative alliances, largely excluding the work of the external participants. On occasion at the behest of the ARL Director, additional reviews are conducted, which may include more comprehensive consideration of one or more of these alliances. During the current review cycle, some of the work of several alliances was reviewed as part of the enterprise reviews of the Network Science and Autonomous Systems programs.

It is becoming increasingly evident that greater attention should be paid to the work done by all participants in collaborative alliances, both intra- and extra-mural. If the alliances are to succeed as intended, then their efforts need to make profound impacts on the content and quality of ARL's portfolio as well as on the accomplishments of its staff. If this is so, then management should wish for validation from external review. To the extent that it is not so, management should welcome the advice and counsel from the external review process. ARL should consider establishing an independent review that will allow for adequate examination of the work done by all parties in the collaborative alliances. Flexibility is the critical guiding principle. No single format or frequency of review is likely to fit all situations.

Impact Analysis

The issue of impact analysis was discussed earlier as it relates generally to all ARL programs, whether carried out within ARL or through extramural effort. When applied to collaborative alliances, this issue takes on special importance. Within the Department of Defense, the Army is unique in its extramural engagement strategy for performance and management of mission-critical S&T. It seems appropriate to ask whether such an approach could benefit the Air Force and Navy, but it is difficult to make the case either for or against. Indeed, it is difficult to find documentation that clarifies the advantages and disadvantages of this approach and its comparative value to the traditional use of government laboratories by the other two services. ARL should consider addressing this concern by performing or commissioning retrospective analyses of these extramural collaborative activities, to be targeted at such issues as best practice in management, technical accomplishments, and impacts on the Army and on the conduct of business in ARL.

Appendixes

Appendix A

Army Research Laboratory Organization Chart

Figure A.1 presents an organization chart of the Army Research Laboratory.

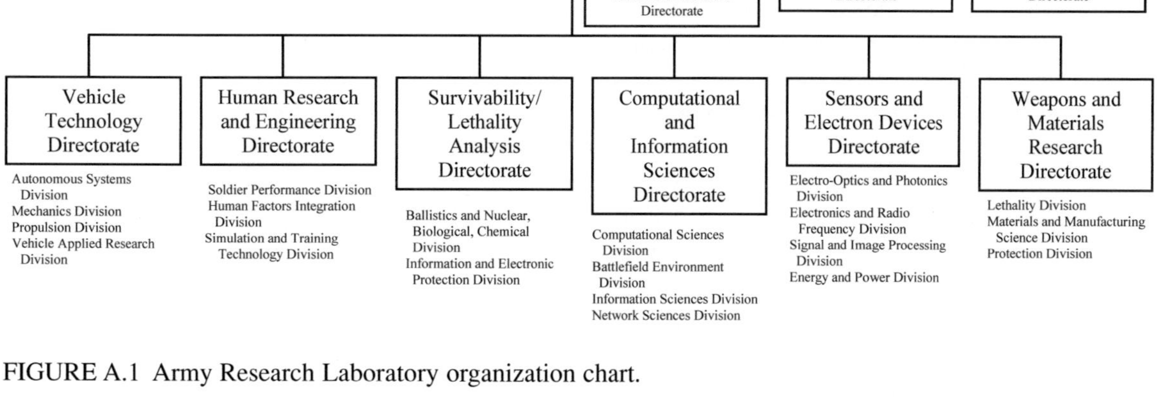

FIGURE A.1 Army Research Laboratory organization chart.

Appendix B

Membership of the Army Research Laboratory Technical Assessment Board and Its Panels

This appendix presents biographical sketches of the members of the Army Research Laboratory Technical Assessment Board, followed by listings of the members of the panels and their affiliations.

BIOGRAPHICAL SKETCHES OF MEMBERS: ARMY RESEARCH LABORATORY TECHNICAL ASSESSMENT BOARD

LYLE H. SCHWARTZ, *Chair,* NAE, is a retired director of the Air Force Office of Scientific Research (AFOSR) and currently a senior research scientist with the Department of Materials Science and Engineering at the University of Maryland. He was a professor of materials science and engineering at Northwestern University for 20 years and the director of Northwestern's Materials Research Center for 5 of those years. He then became the director of the Materials Science and Engineering Laboratory at the National Institute of Standards and Technology, where he served for more than 12 years. His responsibilities included management of the research and development (R&D) agenda in the context of a government laboratory. Dr. Schwartz subsequently assumed responsibility for basic research on structural materials of interest to the U.S. Air Force, in addition to the areas of propulsion, aeromechanics, and aerodynamics. He then completed his government service as director of the AFOSR with responsibility for the entire basic research program of the Air Force. His current interests include government policy for R&D, particularly for materials R&D; materials science education at K-12 levels; and enhanced public understanding of the roles and importance of technology in society. Dr. Schwartz received both his B.S. in engineering and Ph.D. in materials science from Northwestern University.

DONALD M. CHIARULLI is a professor of computer science and computer engineering at the University of Pittsburgh. His expertise includes experimental computer architecture, and optics and optoelectronics for dense interconnection networks. In the context of building experimental systems, his work also includes a significant effort in the development of new design tools for the modeling and simulation of these systems. Dr. Chiarulli holds patents in computer and related optical and optoelectronic hardware. His current research work is in the areas of chip-level optoelectronic interconnections, optical-electronic-mechanical multidomain computer-aided design, optical memory systems, robotics, and voice input/ speech output interfaces for embedded system applications. He received his Ph.D. in computer science from Louisiana State University.

DAVID E. CROW, NAE, is retired senior vice president of engineering at Pratt and Whitney Aircraft Engine Company. He is currently a professor of mechanical engineering at the University of Connecticut. At Pratt and Whitney he was influential in design, development, testing, and manufacturing in support of a full line of engines for aerospace and industrial applications. Dr. Crow was involved with products that include high-thrust turbofans for large commercial and military aircraft, turboprops and small turbofans for regional and corporate aircraft and helicopters, booster engines and upper-stage propulsion systems for advanced launch vehicles, turbopumps for the space shuttle, and industrial engines for land-based power generation. His involvement included sophisticated computer modeling and standards work to bring constant improvements in the performance and reliability of the company's products while at the same time reducing noise and emissions.

MARJORIE ERICKSON is an expert both in the development of physics-based models of material behavior in the prediction of material failure and in the performance of risk assessments. Dr. Erickson is the president of Phoenix Engineering Associates, Inc., and she is an adjunct professor of mechanical engineering at the University of Maryland. She conducts research and consults with industry regarding fracture safety assessment methodology for steel and other alloy components. She provides these services in the areas of assessing the integrity and durability of civil, mechanical, and marine structures fabricated from metallic materials. Specific work that Dr. Erickson has performed includes developing and using integrated, predictive models of material behavior to assess the current status and predict the remaining safe life, under known or expected operating and accident-event conditions, for nuclear pressure vessels and other alloy applications, including fracture safety assessment and life extension of aging aircraft and pipelines. Dr. Erickson received her Ph.D. in materials science from the University of Virginia.

DEBASIS MITRA, NAE, is vice president in the Chief Scientist's Office of Bell Labs, Alcatel-Lucent. He is responsible for global research partnerships and academic relations. From 1999 to 2007 as vice president of the Mathematical and Algorithmic Sciences Research Center, he directed activities in fundamental mathematics, algorithms, complex systems analysis and optimization, statistics, learning theory, information and communications sciences, and industrial mathematics. He is a Bell Labs Fellow and a Life Fellow of the IEEE. He has been McKay Professor at the University of California, Berkeley, and the Albert Winsemius Professor at the Nanyang Technical University in Singapore.

R. BYRON PIPES, NAE, is the John L. Bray Distinguished Professor of Engineering at Purdue University. He is a member of the Royal Society of Engineering Sciences of Sweden (1995). Composite materials have been the focus of his scholarship for the past 28 years. He has developed analytical models and carried out experiments with the objective of developing a fundamental understanding of the design, durability, and manufacturing of these materials systems and structures. He served as Goodyear Endowed Professor of Polymer Engineering at the University of Akron during 2001-2004. He was Distinguished Visiting Scholar at the College of William and Mary during 1999-2001, where he pursued research at the NASA Langley Research Center in the field of carbon nanotechnology. He served as president of Rensselaer Polytechnic Institute from 1993 to 1998. Dr. Pipes was provost and vice president for academic affairs at the University of Delaware from 1991 to 1993 and served as dean of the College of Engineering and director of the Center for Composite Materials during 1977 to 1991 at the same institution. He was appointed Robert L. Spencer Professor of Engineering in 1986 in recognition of his outstanding scholarship in the field of polymer composite materials spanning the subject areas of advanced manufacturing science, durability, design, and characterization. Dr. Pipes received his Ph.D. degree in mechanical engineering from the University of Texas at Arlington and his MSE from Princeton University.

JEREMY M. WOLFE is a professor of ophthalmology and radiology at Harvard Medical School and director of the Visual Attention Laboratory and of the Radiology Department's Center for Advanced Medical Imaging at Brigham and Women's Hospital. In addition, he is a visiting faculty member in the Department of Brain and Cognitive Sciences at the Massachusetts Institute of Technology (MIT) and an adjunct associate professor in cognitive and neural systems at Boston University. He has extensive expertise in vision, binocular perception, visual attention, and cognitive science. Dr. Wolfe has received numerous honors and awards throughout his career and holds memberships in a number of prominent professional societies and organizations. He has authored 112 published papers, 1 textbook, and 26 book chapters. He received his Ph.D. in psychology from MIT.

Staff

JAMES P. McGEE is the director of the Laboratory Assessments Board, the Army Research Laboratory Technical Assessment Board (ARLTAB), and the Committee on National Institute of Standards and Technology Technical Programs in the Division on Engineering and Physical Sciences (DEPS) at the National Research Council (NRC). Since 1994, he has been a senior staff officer at the NRC, directing projects in the areas of systems engineering and applied psychology, including activities of ARLTAB and projects of the Committee on National Statistics' Panel on Operational Testing and Evaluation of the Stryker Vehicle and the Committee on Assessing the National Science Foundation's Scientists and Engineers Statistical Data System, the Committee on the Health and Safety Needs of Older Workers, and the Steering Committee on Differential Susceptibility of Older Persons to Environmental Hazards. He has also served as staff officer for NRC projects on air traffic control automation, musculoskeletal disorders and the workplace, and the changing nature of work. Prior to joining the NRC, Dr. McGee held technical and management positions in systems engineering and applied psychology at IBM, General Electric, RCA, General Dynamics, and United Technologies corporations. He received his B.A. from Princeton University and his Ph.D. from Fordham University, both in psychology, and for several years instructed postsecondary courses in applied psychology and in organizational management.

ARUL MOZHI is senior program officer at the Laboratory Assessments Board. Since 1999, he has been a senior program officer at the NRC, directing projects in the areas of defense science and technology, including those carried out by numerous study committees of the Laboratory Assessments Board, the Army Research Laboratory Technical Assessment Board, the Naval Studies Board, and the National Materials and Manufacturing Board. Prior to joining the NRC, Dr. Mozhi held technical and management positions in systems engineering and applied materials research and development at UTRON, Inc.; Roy F. Weston, Inc.; and Marko Materials, Inc. He received his M.S. and Ph.D. degrees (the latter in 1986) in materials engineering from the Ohio State University and then served as a postdoctoral research associate there. He received his B.S. in metallurgical engineering from the Indian Institute of Technology, Kanpur, in 1982.

LIZA HAMILTON is the administrative coordinator for the Laboratory Assessments Board. Since 2002, she has been responsible for managing the administrative aspects of panel formation, panel meetings, report publication and dissemination, and program development. In addition, she has designed newsletters, brochures, cover designs, and figures for numerous reports prepared by the NRC's Division on Life Sciences and DEPS. Ms. Hamilton earned a 4-year certification in musical theater performance from Pinellas County Center for the Arts in St. Petersburg, Florida; a B.F.A. in film studies from the University of Utah; a design certification from Maryland Institute College of Art; and a master's of liberal arts from Johns Hopkins University.

EVA LABRE is the program associate for the Laboratory Assessments Board. Since 2009, she has been responsible for assisting in the management of the administrative aspects of panel formation, panel meetings, report publication and dissemination, and program development. In addition, she has been responsible for travel expense accounting. Ms. Labre previously held administrative positions on the staff of the Committee on International Organizations and Programs in the NRC Office of International Affairs and on the staff of the Research Associateship Program in the NRC Office of Scientific and Engineering Personnel. Ms. Labre has a B.A. in art history from George Washington University.

ROSE NEUGROSCHEL served as the research associate for the Laboratory Assessments Board until January 20, 2012. She was responsible for the research and security-related tasks of panel formation and report publications, including gathering and evaluating background materials for the committee, assisting panel members in the security clearance process, and ensuring that restricted materials are properly handled. Before joining the Laboratory Assessments Board, Ms. Neugroschel worked as a research assistant for the Board on Testing and Assessment in the Division on Behavioral and Social Sciences and Education at the NRC. She earned a B.A. in psychology from James Madison University.

PANEL ROSTERS

Panel on Air and Ground Vehicle Technology

David Crow, NAE, Pratt and Whitney Aircraft Engine Company (retired), *Chair*
Ralph Aldredge, University of California, Davis
James Bettner, Propulsion Consultant, Pittsboro, Indiana
Paul Bevilaqua, NAE, Lockheed Martin Aeronautics Company
Earl Dowell, NAE, Duke University
Ephrahim Garcia, Cornell University
Prabhat Hajela, Rensselaer Polytechnic Institute
James Hamilton, Target Chip Ganassi Racing
Roger L. McCarthy, NAE, McCarthy Engineering
William Q. Meeker, Jr., Iowa State University
Lynne Parker, University of Tennessee
Neil Paton, NAE, Liquidmetal Technologies
Martin Peryea, Bell Helicopter Textron, Inc.

William Sirignano, NAE, University of California, Irvine
Gaurav S. Sukhatme, University of Southern California
Michael Torok, Sikorsky Aircraft Corporation

Panel on Armor and Armaments

R. Byron Pipes, NAE, Purdue University, *Chair*
Melvin R. Baer, Sandia National Laboratories (retired)
Richard M. Christensen, NAE, Stanford University
Jack J. Dongarra, NAE, University of Tennessee, Knoxville
Thomas Eagar, NAE, Massachusetts Institute of Technology
Mark Eberhart, Colorado School of Mines
Randall M. German, San Diego State University
Michael Jaffe, New Jersey Institute of Technology
Bernard H. Kear, NAE, Rutgers, The State University of New Jersey
Clarence W. "Wes" Kitchens, Jr., Wes Kitchens and Associates, LLC
Kenneth K. Kuo, Pennsylvania State University
Gregory Miller, University of California, Davis
Tresa M. Pollock, NAE, University of California, Santa Barbara
George C. Schatz, NAS, Northwestern University
Steven F. Son, Purdue University
Leonard Uitenham, North Carolina Agricultural and Technical State University

Panel on Digitization and Communications Science

Debasis Mitra, NAE, Bell Labs, Alcatel-Lucent, *Chair*
C. Gordon Bell, NAS/NAE, Microsoft Research
Keren Bergman, Columbia University
Joel S. Birnbaum, NAE, Hewlett-Packard Company (retired)
David Borth, Motorola, Inc.
L. Reginald Brothers, Jr., BAE Systems
Gary Brown, Virginia Polytechnic Institute and State University
George Karypis, University of Minnesota
Christina B. Katsaros, Northwest Research Associates, Inc.
Stephen T. Kent, Raytheon BBN Technologies
Thomas L. Koch, Lehigh University
Juan C. Meza, Lawrence Berkeley National Laboratory
Linda A. Ness, Applied Communication Sciences Applied Research
Tamar Peli, Charles Stark Draper Laboratory, Inc.
Radia Perlman, Intel Labs
Mikel Petty, University of Alabama, Huntsville
Padma Raghavan, Pennsylvania State University
Nelson L. Seaman, Pennsylvania State University
Neil G. Siegel, NAE, Northrop Grumman Information Systems
John Snow, University of Oklahoma
Salvatore J. Stolfo, Columbia University

Panel on Sensors and Electron Devices

Donald Chiarulli, University of Pittsburgh, *Chair*
Eli Brookner, Raytheon Company
Charles R. Cantor, NAS, Sequenom, Inc.
Amy E. Duwel, Charles Stark Draper Laboratory
Thomas Fuller, Georgia Institute of Technology
Elsa M. Garmire, NAE, Dartmouth College
George I. Haddad, NAE, University of Michigan
Herbert Hess, University of Idaho
David A. Hodges, NAE, University of California, Berkeley
Paul Hoff, Independent Consultant, Bedford, New Hampshire
Jennie S. Hwang, NAE, H-Technologies Group
Douglas Mook, The Aptec Group
Randolph L. Moses, Ohio State University
Erin K. O'Shea, NAS, Harvard University
Alan Rudolph, Defense Threat Reduction Agency
Jorge J. Santiago-Aviles, University of Pennsylvania
Michael G. Spencer, Cornell University
Levi Thompson, University of Michigan
Matthew V. Tirrell, NAE, University of Chicago
Anil V. Virkar, University of Utah
Tuan Vo-Dinh, Duke University
Larry P. Walker, Cornell University

Panel on Survivability and Lethality Analysis

Marjorie Erickson, Phoenix Engineering Associates, Inc., *Chair*
David Aucsmith, Microsoft Corporation
Alfred O. Awani, The Boeing Company
David Barton, NAE, Independent Consultant, Hanover, New Hampshire
Gerald G. Brown, NAE, U.S. Naval Postgraduate School
W. Peter Cherry, NAE, Independent Consultant
Ronald R. Luman, Johns Hopkins University
Guruswami Ravichandran, California Institute of Technology
Stephen M. Robinson, NAE, University of Wisconsin-Madison
Armando A. Rodriguez, Arizona State University
Frank J. Serna, Charles Stark Draper Laboratory
Marlin U. Thomas, Air Force Institute of Technology
Donna K. Vargas, Independent Consultant, Las Cruces, New Mexico
Alan R. Washburn, NAE, U.S. Naval Postgraduate School

Soldier Systems Panel

Jeremy Wolfe, Brigham and Women's Hospital and Harvard Medical School, *Chair*
Theodore Berger, University of Southern California

Tora Bikson, The RAND Corporation
Kenneth R. Boff, Independent Consultant
Linda Ng Boyle, University of Washington
Michael Byrne, Rice University
Terry Connolly, University of Arizona
Nancy J. Cooke, Arizona State University
Dennis G. Faust, Lockheed Martin Corporation
J. Dexter Fletcher, Institute for Defense Analyses
Paul W. Glimcher, New York University
Steven A. Hillyard, University of California, San Diego
Verlin B. Hinsz, North Dakota State University
Daniel Ilgen, Michigan State University
Arthur F. Kramer, University of Illinois
Gerald P. Krueger, Krueger Ergonomics Consultants, Alexandria, Virginia
Barbara G. Shinn-Cunningham, Boston University
Milind Tambe, University of Southern California
Charles S. Watson, Indiana University
Holly Yanco, University of Massachusetts, Lowell

Appendix C

Assessment Criteria

The Army Research Laboratory Technical Assessment Board's assessment considered the following general questions posed by the Army Research Laboratory (ARL) Director:

1. Is the scientific quality of the research of comparable technical quality to that executed in leading federal, university, and/or industrial laboratories both nationally and internationally?
2. Does the research program reflect a broad understanding of the underlying science and research-conducted elsewhere?
3. Does the research employ the appropriate laboratory equipment and/or numerical models?
4. Are the qualifications of the research team compatible with the research challenge?
5. Are the facilities and laboratory equipment state of the art?
6. Does the research reflect an understanding of the Army's requirement for the research or the analysis?
7. Are programs crafted to employ the appropriate mix of theory, computation, and experimentation?
8. Is the work appropriate to the ARL niche?
9. Are there especially promising projects that, with application of adequate resources, could produce outstanding results that could be transitioned ultimately to the field?

The Board applied the following metrics or criteria to the assessment of the scientific and technical work reviewed at ARL:

1. Effectiveness of Interaction with the Scientific and Technical Community
   a. Papers in quality refereed journals and conference proceedings (and their citation index)
   b. Presentations and colloquia
   c. Participation in professional activities (society officers, conference committees, journal editors)

d. Educational outreach (serving on graduate committees, teaching or lecturing, invited talks, mentoring students)
e. Fellowships and awards (external and internal)
f. Review panel participation (Army Research Office, National Science Foundation, Multidisciplinary University Research Initiative)
g. Recruiting new talent into the ARL
h. Patents and intellectual property (IP) (and examples of how the patent or IP is used)
i. Involvement in building an ARL-wide cross-directorate community
j. Public recognition (e.g., in the press and elsewhere) for ARL research

2. Impact on Customers
   a. Documented transfer or transition of technology, concepts, or program assistance from ARL to Research, Development, and Engineering Centers (RDECs) or RDEC contractors for both the long term and short term
   b. Direct funding from customers to support ARL activities
   c. Documented demand for ARL support or services (is there competition for ARL's support?)
   d. Customer involvement in directorate planning
   e. Participation in multidisciplinary, cross-directorate projects
   f. Surveys of customer base (direct information from customers on value of ARL research)

3. Formulation of Projects' Goals and Plans
   a. Is there a clear tie to ARL Strategic Focus Areas, Strategic Plan, or other ARL need?
   b. Are tasks well defined to achieve objectives?
   c. Does the project plan clearly identify dependencies (i.e., successes depend on success of other activities within the project or outside developments)?
   d. If the project is part of a wider activity, is role of the investigators clear, and are the project tasks and objectives clearly linked to those of other related projects?
   e. Are milestones identified if they are appropriate? Do they appear feasible?
   f. Are obstacles and challenges defined (technical, resources)?
   g. Does the project represent an area where application of ARL strengths is appropriate?

4. Research and Development Methodology
   a. Are the hypotheses appropriately framed within the literature and theoretical context?
   b. Is there a clearly identified and appropriate process for performing required analyses, prototypes, models, simulations, tests, etc.?
   c. Are the methods (e.g., laboratory experiment, modeling or simulation, field test, analysis) appropriate to the problems? Do these methods integrate?
   d. Is the choice of equipment or apparatus appropriate?
   e. Is the data collection and analysis methodology appropriate?
   f. Are conclusions supported by the results?
   g. Are proposed ideas for further study reasonable?
   h. Do the trade-offs between risk and potential gain appear reasonable?
   i. If the project demands technological or technical innovation, is that occurring?
   j. What stopping rules, if any, are being or should be applied?

5. Capabilities and Resources
   a. Are the qualifications and number of the staff (scientific, technical, administrative) appropriate to achieve success of the project?
   b. Is funding adequate to achieve success of the project?
   c. Is the state of the equipment and facilities adequate?
   d. If staff, funding, or equipment is not adequate, how might the project be triaged (what thrust should be emphasized, what sacrificed?) to best move toward its stated objectives?
   e. Does the laboratory sustain the technical capability to respond quickly to critical issues as they arise?

6. Responsiveness to the Board's Recommendations
   a. Have the issues and recommendations presented in the previous report been addressed?

Appendix D

Acronyms and Abbreviations

ACM	Association for Computing Machinery
AEC	Army Evaluation Center
AFOSR	Air Force Office of Scientific Research
AMRDEC	Aviation and Missile Research, Development and Engineering Center
AMSAA	Army Materiel Systems Analysis Activity
APS	active protection system
ARL	Army Research Laboratory
ARLTAB	Army Research Laboratory Technical Assessment Board
ARO	Army Research Office
ATO	Army Technology Objective
BED	Battlefield Environment Division
BRAC	base realignment and closure
CECOM	Communications-Electronics Command
CFD	computational fluid dynamics
CISD	Computational and Information Sciences Directorate
CMD-P	Computer Meteorological Data Profiler
CMMI	capability maturity model integration
CRA	Collaborative Research Alliance
CSD	Computational Sciences Division of the CISD
CTA	Collaborative Technology Alliance

DAC	digital-to-analog converter
DARPA	Defense Advanced Research Projects Agency
DE	directed energy
DFT	density functional theory
DMAZ	dimethylamino-2-ethylazide
DNS	direct numerical simulation
DoD	Department of Defense
DOE	Department of Energy
DRI	Director's Research Initiative
DSI	Director's Strategic Initiative
EAR	Environment for Auditory Research
EEG	electroencephalogram
EMVAF	Electro Magnetic Vulnerability Assessment Facility
EO&P	Electro-optics and Photonics (Division)
FCS	Future Combat Systems
GCV	Ground Combat Vehicle
GNC	guidance, navigation, and control
GPS	Global Positioning System
GSR	galvanic skin response
hp	horsepower
HPC	high-performance computing
HRED	Human Research and Engineering Directorate
HRI	Human-robot interaction
HSI	Human-systems integration
IED	improvised explosive device
IEEE	Institute of Electrical and Electronics Engineers
IMPRINT	Improved Performance Research Integration Tool (software)
INVA/DE	Integrated Network Vulnerability Assessment/ Discovery Exploitation
IPA	Intergovernmental Personnel Act
IR	infrared
IRCM	infrared countermeasures
IRFNA	inhibited red fuming nitric acid
ISD	Information Sciences Division
ISR	intelligence, surveillance, and reconnaissance
ITA	International Technology Alliance
JP-8	Jet Propellant 8 fuel
KE	kinetic energy
LED	light-emitting diode
LES	large-eddy simulation

LIF	laser-induced fluorescence
Li-ion	lithium-ion
MADIS	Meteorological Assimilation Data Ingest System
MANPRINT	Manpower and Personnel Integration
MAST	Micro Autonomous Systems and Technology
MBT&E	Mission-Based Test and Evaluation
MEDE	Materials in Extreme Dynamic Environments
MEMS	microelectromechanical systems
MMF	Mission and Means Framework (software)
MSME	Multiscale Multidisciplinary Modeling of Electronic Materials
MURI	Multidisciplinary University Research Initiative
MUVES	Modular UNIX-based Vulnerability Estimation Suite (software)
NASA	National Aeronautics and Space Administration
NLOS	non-line-of-sight
NOAA	National Oceanographic and Atmospheric Administration
NRC	National Research Council
NSD	Network Sciences Division
NSF	National Science Foundation
NWS	National Weather Service
OA	optical augmentation
ODR	opportunity-driven research
PSL	New Mexico State University Physical Sciences Laboratory
QWIP	quantum-well infrared photodetector
R&D	research and development
RDEC	Research, Development, and Engineering Center
RF	radio frequency
RTDS	real-time digital simulator
S&T	science and technology
S4	System-of-Systems Survivability Simulation (software)
SBIR	Small Business Innovation Research
SEDD	Sensors and Electron Devices Directorate
SEMACS	Specialty Electronics, Materials, and Sensors (Program)
SiC	silicon carbide
SLAD	Survivability/Lethality Analysis Directorate
SLAM	simultaneous localization and mapping
SLV	survivability, lethality, and vulnerability
SME	subject matter expert
SoS	system-of-systems
SoSA	system-of-systems analysis
SPEAR	Soldier Performance and Equipment Advanced Research

STTC	Simulation and Training Technology Center
SWaP	size, weight, and power
TARDEC	Tank Automotive Research, Development, and Engineering Center
TBI	traumatic brain injury
T&E	testing and evaluation
TMEDA	N,N,N′,N′-tetramethylethylenediamine
TRAC	Training and Doctrine Command Analysis Center
TRADOC	Training and Doctrine Command
UAV	unmanned aerial vehicle
UBB	underbody blast
UGV	unmanned ground vehicle
UV	ultraviolet
VARD	Vehicle Applied Research Division
VTD	Vehicle Technology Directorate
VTOL	vertical takeoff and landing
WIAMan	Warrior Injury Assessment Manikin
WMRD	Weapons and Materials Research Directorate
WRF	weather research and forecast